SpringerBriefs in Applied Sciences and Technology

For further volumes:
http://www.springer.com/series/8884

Fan Yang · Ping Duan · Sirish L. Shah
Tongwen Chen

Capturing Connectivity and Causality in Complex Industrial Processes

 Springer

Fan Yang
Tsinghua Laboratory for Information
  Science and Technology
Department of Automation
Tsinghua University
Beijing
China

Ping Duan
Tongwen Chen
Department of Electrical and Computer
  Engineering
University of Alberta
Edmonton, AB
Canada

Sirish L. Shah
Department of Chemical and Materials
  Engineering
University of Alberta
Edmonton, AB
Canada

ISSN 2191-530X   ISSN 2191-5318   (electronic)
ISBN 978-3-319-05379-0   ISBN 978-3-319-05380-6   (eBook)
DOI 10.1007/978-3-319-05380-6
Springer Cham Heidelberg New York Dordrecht London

Library of Congress Control Number: 2014934663

Printed on acid-free paper

Springer is part of Springer Science+Business Media (www.springer.com)

To our families

Preface

Large-scale complex systems, such as modern industrial processes, biological systems, and social networks, are interconnected by different units or elements; the system behavior is determined by the inter-relationship between every pair of the elements as well as the local dynamics within each element. It is essential to identify such inter-relationship, namely connectivity and causality, in order to analyze influence mechanisms, structural properties, and overall dynamic behavior.

In the control and automation community, connectivity and causality play a vital role in modeling and analysis, especially for fault detection and hazard analysis, because an abnormality can easily propagate within and between process units due to material and information flow paths. Thus the problem of fault detection and isolation for industrial processes is concerned with determination of root causes and fault propagation. Connectivity and causality, as the key features of process description, can be captured in two ways:

1. From process knowledge: Structural modeling based on first principles structural models can be merged with adjacency/reachability matrices or topology models obtained from process flow-sheets described in standard formats.
2. From process data: Cross-correlation analysis, Granger causality and its extensions, frequency domain methods, information-theoretic methods, and Bayesian networks can be used to identify pairwise relationship and network topology.

These methods rely on the notion of information fusion, whereby various types of process operating data are combined with qualitative process knowledge to give a holistic picture of the system.

In this book, we shall give an exhaustive overview of concepts and descriptions of connectivity and causality in complex processes and a tutorial guide to classical and recent research results on detection of connectivity and causality illustrated with example applications. A study of the fusion of different information resources for obtaining an acceptable process topology is also introduced.

Some details are omitted in this book due to space constraints. Interested readers should refer to the related literature. For questions, comments, and suggestions, please write to Fan Yang at yangfan@tsinghua.edu.cn.

December 2013   Fan Yang
                                                                                  Ping Duan
                                                                             Sirish L. Shah
                                                                             Tongwen Chen

Acknowledgments

This book would not have been possible without the help from many people.

Fan Yang expresses his sincere thanks to Prof. Deyun Xiao at Tsinghua University (Beijing, China) for his long-term guidance and help. Fan Yang's student, Mr. Cen Guo, kindly helped in editing some part of the draft. Fan Yang is grateful for financial support from the National Natural Science Foundation of China (No. 60904044), the National High-tech R&D Program of China (No. 2013AA040702) and the Foundation of Key Laboratory of Advanced Process Control for Light Industry (Jiangnan University), Ministry of Education, China.

Ping Duan is thankful to Prof. Nina F. Thornhill from Imperial College London, UK, for comments and suggestions on her research work on direct causality analysis, and to Dr. Yuri Shardt for help with the three-tank experiment.

Sirish L. Shah thanks his students, research fellows, and his grandchildren (Maya, Keshav, Anuva, and Avni), all of whom have been the main sources or CAUSES of his enriched life.

Tongwen Chen thanks his past post-doctoral fellows and graduate students who collaborated on related research topics on advanced alarm monitoring, including Iman Izadi, Jiandong Wang, Yue Cheng, Naseeb Adnan, Kabir Ahmed, and Md Shahedul Amin. Tongwen Chen's research in this area was supported by a strategic project grant from the Natural Sciences and Engineering Research Council of Canada entitled "Development of an Advanced Technology for Alarm Analysis and Design" (2009–2012).

Acknowledgment is given to the Institute of Electrical and Electronic Engineers (IEEE) to reproduce material from the following papers:

©2013 IEEE. Reprinted, with permission, from Duan P, Yang F, Chen T, Shah SL (2013) Direct causality detection via the transfer entropy approach. IEEE Transactions on Control Systems Technology, 21(6):2052–2066 (as related to material in Chap. 5)

©2009 IEEE. Reprinted, with permission, from Yang F, Shah SL, Xiao D (2009) SDG model-based analysis of fault propagation in control systems. In: Proc. 2009 Canadian Conference on Electrical and Computer Engineering, St John's, NL, Canada, pp 1152–1157 (as related to material in Chap. 4)

Acknowledgment is given to International Federation of Automatic Control (IFAC) to reproduce material from the following paper:

©2010 IFAC. Originally published in IFAC-PapersOnLine. Reprinted, with permission, from Yang F, Shah SL, Xiao D (2010) signed directed graph (SDG)-based process description and fault propagation analysis for a tailings pumping process. In: Proceedings of 13th Symposium on Automation in Mining, Mineral and Metal Processing, Cape Town, South Africa, pp 50–55 (as related to material in Chap. 6)

Acknowledgment is given to Elsevier to reproduce material from the following paper:

©2006 Elsevier. Reprinted, with permission, from Yim SY, Ananthakumar HG, Benabbas L, Horch A, Drath R, Thornhill N (2006) Using process topology in plant-wide control loop performance assessment. Computers and Chemical Engineering, 31(2):86–99 (Fig. 4.1 in Chap. 4)

Acknowledgment is given to University of Zielona Góra and Lubuskie Scientific Society to reproduce material from the following paper:

©2011 University of Zielona Góra and Lubuskie Scientific Society. Reprinted, with permission, from Yang F, Shah SL, Xiao D (2011) Signed directed graph-based modeling and its validation from process knowledge and process data. International Journal of Applied Mathematics and Computer Science, 22(1):41–53 (as related to material in Chaps. 5 and 6)

Contents

1   Introduction 1
   1.1   Concept of Causality 2
   1.2   Connectivity, Correlation, and Causality................. 4
   1.3   Preview of Chapters............................... 5
   1.4   Chapter Summary 6
   References 6

2   Examples of Applications for Connectivity
and Causality Analysis................................. 7
   2.1   Topology Modeling and Closed-Loop Identification 7
   2.2   Root Cause Analysis 8
   2.3   Risk Analysis: HAZOP 9
   2.4   Consequential Alarm Identification 9
   2.5   Plant-Wide Control Structure Design 10
   2.6   Chapter Summary 10
   References 11

3   Description of Connectivity and Causality................... 13
   3.1   Description of Connectivity 13
      3.1.1   Adjacency Matrices......................... 14
      3.1.2   Digraphs 14
      3.1.3   Semantic Web Description..................... 16
   3.2   Description of Causality............................ 18
      3.2.1   Structural Equation Models 18
      3.2.2   Matrices and Digraphs........................ 20
      3.2.3   Matrix Layout Plots 21
   3.3   Chapter Summary 22
   References 22

**4   Capturing Connectivity and Causality
from Process Knowledge**   23
   4.1   Structural Modeling Based on First-Principle
Structural Models   24
   4.2   Construction of Adjacency and Reachability Matrices   24
   4.3   Construction of Graphical Models......................   24
      4.3.1   Modeling of SDGs   24
      4.3.2   SDG Modeling in Control Systems   29
      4.3.3   Other Graphical Models.......................   35
   4.4   Rule-Based Models   36
   4.5   Extracting Plant Topology from Web Language   37
   4.6   Chapter Summary   38
   References   38

5   Capturing Causality from Process Data   41
   5.1   System Identification Approach.........................   42
   5.2   Cross-Correlation Analysis   45
   5.3   Granger Causality Analysis...........................   47
   5.4   Directed Transfer Function/Partial Directed
Coherence Analysis   51
   5.5   Transfer Entropy Analysis   54
   5.6   Bayesian Network Learning   58
   5.7   Other Methods   60
   5.8   Mutual Validation by Process Knowledge and Data..........   60
      5.8.1   Using Process Data to Validate
Knowledge Description   60
      5.8.2   Using Process Knowledge to Validate
Data-Based Relations.........................   62
   5.9   Chapter Summary   63
   References   63

6   Case Studies   67
   6.1   Three-Tank System   67
      6.1.1   Adjacency Matrix Method   68
      6.1.2   Granger Causality Method   70
      6.1.3   Transfer Entropy Method.......................   71
      6.1.4   Bayesian Network Structure Inference Method   72
   6.2   Eastman Process   72
      6.2.1   Adjacency Matrix Method   75
      6.2.2   Data Driven Methods   77

6.3 Final Tailings Pump House Process.................... 82
  6.3.1 Process Description.......................... 83
  6.3.2 Using Process Data to Validate Knowledge
         Description............................... 84
  6.3.3 Using Process Knowledge to Validate
         Data-Based Relations........................ 87
  6.3.4 Application of SDGs in Fault Propagation Analysis 88
6.4 Chapter Summary 88
References 89

Glossary 91

Chapter 1
Introduction

Abstract In large-scale industrial processes and other complex systems, elements are not independent. To describe the relationship between process variables, different concepts, such as connectivity and causality, are often used. The background and motivation of investigatingconnectivity and causality in complex systems are discussed; these two concepts are clarified with examples. Causality describes the cause-effect relationship between changes of process variables; while connectivity is generally concerned with physical and information paths in a process. The causality relationships can be described by process topology. A brief chapter preview is then included to give a big picture of this brief, and to provide guidance for interested readers to different topics covered in the brief.

Keywords Complex processes · Process variables · Causality · Connectivity · Correlation · Predictability · Directionality · Reachability · Process topology

The term causality has different connotations depending on its use in engineering or medicine or philosophy. This aspect is well summarized in the following excerpt from Wikipedia:

"Causality (also referred to as causation) is the relation between an event (the *cause*) and a second event (the *effect*), where the second event is understood as a consequence of the first.

In common usage, causality is also the relation between a set of factors (causes) and a phenomenon (the effect). Anything that affects an effect is a factor of that effect. A direct factor is a factor that affects an effect directly, that is, without any intervening factors. (Intervening factors are sometimes called 'intermediate factors'.) The connection between a cause(s) and an effect in this way can also be referred to as a *causal nexus*.

F. Yang et al., *Capturing Connectivity and Causality in Complex Industrial Processes*, SpringerBriefs in Applied Sciences and Technology, DOI: 10.1007/978-3-319-05380-6_1, © The Author(s) 2014

> Though the causes and effects are typically related to changes or events, candidates include objects, processes, properties, variables, facts, and states of affairs; characterizing the causal relation can be the subject of much debate.
>
> The philosophical treatment on the subject of causality extends over millennia. In the Western philosophical tradition, discussion stretches back at least to Aristotle, and the topic remains a staple in contemporary philosophy."

Connectivity on the other hand has different meaning in the context of mathematics and computer science. It is related to graph theory, and more formally it is defined in terms of connections between nodes or edges. In the context of process topology, which is the main focus of the discussion here, it simply relates to material and information flow connections between or within process units or sensors or actuators or controllers. Thus connectivity is generally concerned with the physical and information paths in a process whether they are direct or indirect. Most notably, the direction of material or information paths or the "arrow of time" is not of interest in establishing connectivity. On the other hand the temporal direction of causation is of critical importance in causality analysis. This is also related to the second law of thermodynamics which says that the "sum of effects can never have lower entropy than the sum of causes". It is this requirement of temporal 'asymmetry' in causality detection which is the key difference between connectivity and causality.

The following example, as shown in Fig. 1.1a, illustrates the main differences between these two similar concepts. 'Pipe 1' (with flow rate F_1) is connected to 'Tank' and 'Tank' is connected to 'Pipe 2' (with flow rate F_2); this is the flow path, where the liquid is fed into the tank and then flow out. Valve 'V' is connected to 'Pipe 2' to control the flow rate by the valve opening. In terms of the information flow path, the signal line is connected to 'V' to transmit the level signal L to the valve. This connectivity is shown in Fig. 1.1b.

Connectivity is not limited to only industrial processes involving material and information transfer. In other types of systems, connectivity may have different meanings. For example, in neurosciences, neurons are connected by synapses due to which electrical or chemical signals can be propagated. However, such propagation is bidirectional. In summary, connectivity shows the material, energy, and information propagation among individual elements and determines the topology of the system.

1.1 Concept of Causality

Elements in a system are not only connected to each other, they are mutually dependent. To describe this dependency, we consider the cause-effect relationships between them, namely, causality (also referred to as causation). Causality describes

Fig. 1.1 Tank example.
a Schematic; **b** Connectivity; **c** Causality

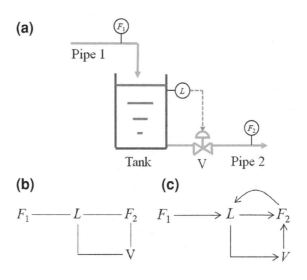

the relationship between events or variables. Causality cannot exist without any
connectivity.

Recall the tank system example in Fig. 1.1a. This process follows the principle of
fluid dynamics; thus F_1 influences L and L influences F_2, the same as in a connectivity
graph, yet F_2 also influences L, which is different from connectivity. For valve 'V',
we are concerned with its stem position, which determines the flow rate F_2 in 'Pipe
2' and is controlled by the level L, transmitted by the signal line. This causality is
shown in Fig. 1.1c.

Intuitively, causality is part of our daily life; however, causality was not accepted
as a scientific concept until statisticians formulated the concept of a randomized
experiment to test causal relations from data [1]. The notion of causality is now
widely accepted as an independent concept and differs from other concepts such
as correlation; correlation does not imply causality. However, an explicit and exact
definition of causality is still difficult. We can give a non-exhaustive list of necessary
(but not sufficient) conditions to establish causality between two variables, X and
Y [2]. A necessary condition for causality requires that between variables x and y
there be a:

(1) a theoretical or common sense linkage;
(2) empirical association (correlation);
(3) elimination of common causes: some other variable must be ruled out as a cause
of the correlation;
(4) responsiveness: altering X leads to an alteration in Y; and
(5) asymmetry: X must cause Y, and not vice vera.

Non-experimental studies can only address the first two conditions and partially
address the third one; whereas others have to be tested by statistical data analysis

in the temporal domain. For this reason, time series analysis should be taken into account to define causality.

Norbert Wiener was one of the first engineers to concieve a mathematical definition of causality: "*X could be termed as to 'cause' Y if the predictability of Y is improved by incorporating information about X*" [3]. Clive Granger adapted this definition into a practical form: "*we say that $x(k)$ is causing $y(k)$ if we are better able to predict $y(k)$ using all available information than if the information apart from $x(k)$ had been used*" [4]. Although this definition does not explain all forms of causality, it has practical utility and thus has been widely known as "Granger causality".

There are other definitions of causality. Another important definition was proposed by Judea Pearl in the probability framework [5]. We will not expand the details of these definitions but focus on the physical meaning and practical use. Note that the concept of causal models in control theory is a different concept; it only reflects the physical realizability of models.

1.2 Connectivity, Correlation, and Causality

In the above list of necessary conditions for causality, the first condition is connectivity, and the second one is correlation. Thus we know that the concept of causality is based upon these two concepts and yet has additional attributes.

The difference between connectivity and causality can be illustrated by the following example: Two tanks can be connected via a pipe with a closed valve on the connecting pipe. Under this condition, the levels in both tanks are noncausal or completely independent of each other. Even if the valve is open, the levels may be noncausal because there may be a control strategy to maintain the levels so that the levels do not show correlation. Even if there is no valve on the connecting pipe, other streams that feed fluid into the tanks and discharge fluid from the tanks can affect the levels so that the levels may not show correlation. Therefore, causality cannot be confirmed without analyzing the data.

Correlation does not imply causality either [6]. A typical example is that x and y are both effects of a third variable z, as shown in Fig. 1.2a. Of course they are correlated, which can be easily tested by investigating the data. However, there is no causality between x and y because if we rule out their common cause z, x and y are both independent. This can be explained by the third condition in the above list and the third variable is called a confounding variable. The identification and treatment of confounding variables is a well-known problem in causality analysis.

Now consider another case: if x causes y and y causes z, then x causes z, as shown in Fig. 1.2b. However, there is a significant difference between correlation and causality: we can say x causes y directly, which means that there is no other variable(s) between them, and x causes z indirectly, which means the causality is based on an intermediate variable y in this case; correlation has no such direct/indirect properties and can be tested for any pair of variables. Another difference between correlation and causality is directionality. Causality has directionality while correlation does not.

Fig. 1.2 Simple topology of three variables

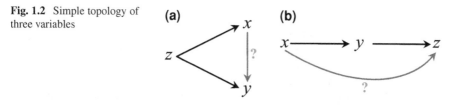

This is reflected in the fifth condition in the above list. One can say x is correlated with y, which is equivalent to saying that y is correlated with x. Whereas if x causes y, one cannot conclude that y causes x. Connectivity can be directed or undirected; it depends upon what is transferred in the connected path.

If we only study the relationship between two variables, the confounding problem is irrelevant. What makes the problem difficult is the multivariate case with a large number of elements or variables; to describe all the relations among them, we construct a network with nodes representing elements or variables and branches denoting their relations; we call this a topology. When we visualize a topology, we may find it very complex, but it illustrates the internal qualitative structure of the system and thus provides useful information for further studies and applications.

By taking into account the directionality, another concept is often useful for topology based on connectivity or causality, namely, reachability [7], to include both direct and indirect relations.

1.3 Preview of Chapters

In this chapter, we have briefly introduced some familiar and somewhat similar and often confusing concepts, particularly connectivity and causality, and thus clarified the research objective of this brief.

There are many areas that are related to the causality analysis. We will briefly discuss several potential application areas in Chap. 2. In fact, the qualitative causality information can assist in control design, fault diagnosis, alarm design and other issues for large-scale complex systems.

Chapter 3 is dedicated to the description of connectivity and causality by mathematical, graphical models, and ontological models.

Chapters 4 and 5, discuss how to capture connectivity and causality. For this purpose, we usually have two resources that can be utilized. In Chap. 4, process knowledge is considered. Based on first-principle structural models, we can obtain equations to describe the structures of systems. The structural models can also be described by qualitative matrices, namely, adjacency and reachability matrices, and therefore the reasoning can be conducted by matrix computation. Apart from mathematical descriptions, semantic web languages, as descriptions of human knowledge, can also be used to describe process knowledge. From the extensible markup language

(XML) or resource description framework/web ontology language (RDF/OWL) files, one can extract topology information with the help of inference engines.

In Chap. 5, another resource, process data, is considered. There are quite a few data-based methods in this category [8]. They were proposed initially in the bivariate case to identify the cause and effect and then extended to the multivariate cases to construct the topology. Thus, in terms of the general problems of identifying pairwise directionality and network topology determination, several common methods are introduced. The simplest one is the cross-correlation analysis to compute the lag-adjusted cross-correlation between every two variables. The temporal order is considered, yet it cannot be extended to multivariate cases without an efficient way to rule out confounding variables. Granger causality analysis is a widely used method in neurosciences; it has some variants to adapt to nonlinear cases. Directed transfer functions (DTF) and partial directed coherence (PDC) are typical frequency domain methods. Based on information theory, transfer entropy can be used to measure the mutual information transferred from one time-series to another; it is a general method and does not need the linearity assumption. From probability theory, conditional expectation is represented by the Bayesian chain rule according to the topology information; thus a Bayesian network can also help us understand causality.

Simulation and real processes are taken as case studies in Chap. 6 where different methods are illustrated and compared.

1.4 Chapter Summary

In this chapter, we have introduced the background and requirements of investigating connectivity and causality in complex systems. These two concepts are clarified and discussed. A brief chapter preview is included to give the big picture of this brief, and to provide guidance for interested readers to various sections or chapters according to their interests and needs.

References

1. Fisher RA (1935) The design of experiments. Oliver and Boyd, Edinburgh
2. Agresti A, Finlay B (2009) Statistical methods for the social sciences, 4th edn. Pearson Prentice Hall, Upper Saddle River
3. Wiener N (1956) The theory of prediction. Modern Mathematics for Engineers, McGraw-Hill, New York, chap 8, pp 165–190
4. Granger CWJ (1969) Investigating causal relations by econometric models and cross-spectral methods. Econometrica 37(3):424–438
5. Pearl J (2009) Causality: models, reasoning, and inference, 2nd edn. Cambridge University Press, Cambridge
6. Wright S (1921) Correlation and causation. J Agric Res 20:557–585
7. Mah RSH (1989) Chemical process structures and information flows. Butterworth, Boston
8. Smith SM, Miller KL, Salimi-Khorshidi G, Webster M, Beckmann CF, Nichols TE, Ramsey JD, Woolrich MW (2011) Network modelling methods for FMRI. NeuroImage 54:875–891

Chapter 2
Examples of Applications for Connectivity and Causality Analysis

Abstract Connectivity and causality have a lot of potential applications, among which we focus on analysis and design of large-scale complex industrial processes. A direct application by establishing connectivity and causality is to build a topological model before parameter identification for complex industrial processes that areusually multi-input, multi-output systems with many internal closed loops. In abnormal situation management, process topology can be employed for root cause analysis, risk analysis, and consequential alarm identification using the information of fault propagation. These potential applications include both off-line analysis andon-line diagnosis. In addition, process topology can eventually be used in design of control structures because process topology determinesthe natural structure of the distributed plant-wide control.

Keywords System identification · Closed loop · Vector autoregressive models · Root cause analysis · Hazard and operabilitystudy · Consequential alarms · Abnormal situation · Plant-wide control structures · Process flow diagrams · Sensor location

Among various applications of connectivity and causality, we will focus on analysis and design of large-scale complex industrial processes. Existing studies have shown great potential of applying connectivity and causality analysis to such cases; the illustration of these applications will also highlight different approaches for causality identification and analysis.

2.1 Topology Modeling and Closed-Loop Identification

A direct application by establishing connectivity and causality is to build a topology model for a complex industrial process. Given process data, system identification is the typical black-box modeling approach. If a known system structure is assumed, there are plenty of methods to estimate parameters. However, in multivariate cases, structure identification is more important and should be performed before parameter estimation. In particular, what we mean by structure here is not only the orders of the local linear models, but also the linkage between variables.

When there are many process variables, it is unwise to separate them into inputs and outputs, because each of them reveals some information of the system and can thus be regarded as a state variable. As a result, traditional quantitative system identification techniques do not work well, and one way is to assume the topology before estimating the orders and parameters of each closed-loop (bidirectional) path model, namely, the local linear model between every two variables. However, there are too many combinations according to the existence of a link between every two variables. Thus, it is more reasonable to assume that each pair of variables are linked together and then estimate the orders and parameters of the path model; if the results show that the link is too weak, then such a link can be removed [9, 10]. This only requires a single-input-single-output (SISO) framework to deal with every path. However for a rigorous analysis, a multiple-input-multiple-output (MIMO) framework, such as a vector autoregressive (VAR) model, should be considered because every variable in a multivariate system may influence as well as be influenced by more than one variable. The above idea suffers from a high computational burden, yet if the topology is known a priori, then the computational burden can be lowered significantly. For this purpose, topology modeling based on process connectivity capturing or process data analytics would help.

2.2 Root Cause Analysis

When the system encounters an abnormal situation, there must be one or more elements showing abnormal symptoms or measurements. If there is only one abnormal element, then this is a local fault in most cases and one should then look into the commensurate part to figure out the problem. If there are multiple abnormal elements, we should be aware that this could be due to some interaction that results in propagation of the source fault. For example, in a pipe network, if an upstream valve is partially blocked, then there will be a series of abnormal events downstream, e.g., reduction of flow rate, decrease of liquid level, and even dry-out of a vessel. When an operator finds that there is something wrong in such a process, there may exist multiple abnormal symptoms; to resolve this situation, the operator should not just tune the valves associated with the vessel for example, because this may make the situation even worse; instead he or she should find the root cause promptly and eliminate it. Once the root cause is resolved, all the other issues disappear accordingly.

Given the topology, or connectivity/causality to be specific, a backward traversal along the paths can be performed to find the root cause, namely, the original abnormal element that causes all the other abnormal elements [17]. What we assume here is that the fault should propagate along the established paths; this is the case most of the times. Among other events the abnormal situations considered in the examples here and generally include cases such as deviation from normal values, oscillations, sensor or actuator malfunction, process or equipment failure, and misoperation.

Take oscillating variables as an example, which is a typical plant-wide disturbance. By using some data-driven methods, oscillating variables can be identified, which are also called efficient nodes in the terminology of signed digraphs because they are the nodes that should be studied. Jiang et al. [11] used a control loop digraph to describe the topology of control loops and, by examining the domain of influence of each control loop were able to find a ranked list of root cause variables to be those that are able to reach all the other oscillating variables along the paths. For a survey of this application, please refer to [5]. Similar work has also been reported in the early study of [4].

2.3 Risk Analysis: HAZOP

Risk analysis is a way to examine a process to identify and evaluate problems that may represent risks. As a representative qualitative approach, hazard and operability study (HAZOP) is frequently applied to planned or existing processes in a structured and systematic way. This task is carried out based on guide words by a series of team meetings. If the topology is available, then this procedure can be relatively straight forward and clear. There are several other studies that use signed digraphs or other graph models for HAZOP study [14–16, 19]. In [20], HAZOP is considered as one of the two main areas of the signed digraph technology (fault diagnosis as another one, as mentioned in the previous section) by using the inference engine essentially based on the search of process topology. Different from root cause analysis, such search is a forward search to find the resulting consequence while the former is a backward search to find the root. The purpose of HAZOP analysis is to find all possible consequences of any assumed faults. But if one wants to estimate the probability of events, quantitative information needs to be incorporated. With such a scheme one can obtain a computer-aided HAZOP analysis.

2.4 Consequential Alarm Identification

Alarm management is an emerging area in the process control community [8]. For monitoring of complex industrial processes, a lot of alarms tags are configured for all kinds of variables. For example, a process variable can trigger high/low alarms to reflect its states. During abnormal situations, alarms should be raised to remind operators to take actions. Ideally, one abnormal situation should trigger one and only one alarm; however, because of redundancy, interactions and correlations between variables even a single abnormal event will result in the annunciation of many alarms. In addition, since a fault can propagate throughout the process, alarms also show up in a specific order. This list of consecutive alarms may be dependent; thus we call them consequential alarms [13]. Consequential alarms over a very short period of time often lead to are construed as alarm floods, leading to a dangerous situation as the console operators or engineers may not be able to identify true root causes. For this case, process topology would be of great help in describing the relationship between alarm tags.

For on-line analysis, this is similar to the root cause analysis because the most important task is to find the root cause of all related or consequential alarms. If the root cause is resolved, then all the alarms can be removed. For off-line analysis, more can be done, for example, to obtain and analyze alarm sequences of typical abnormal situations [3, 12]. When an alarm flood occurs, the alarm sequence is recorded and compared to recorded known sequences to find the most possible root causes so that the previously known and successful alarm mitigation solution can be retrieved immediately; this approach can also be adapted for on-line applications.

2.5 Plant-Wide Control Structure Design

Connectivity and causality reflect the essential nature of a process, so the above applications are all based on a given topology and are aimed at analysis. Such topology can eventually be used in design of control structures because process topology determines the natural structure of the distributed plant-wide control. There are a few studies in this area: Alabi used process flow diagrams (PFDs) in degrees of freedom (DOF) analysis [1]; Cameron and Hangos discussed observability and controllability studies based on structural information [6]; and Hangos and Tuza used graph-theoretical models in optimal control structure selection [7]. These applications of topology can serve as a precursor for other complex and quantitative applications.

Another application of process topology is its use in sensor location. For example, graph models have been used to design feasible and optimal sensor location strategies according to fault detectability and identifiability criteria [2, 18].

2.6 Chapter Summary

Several potential applications of process topology/connectivity and causality have been introduced in this chapter, including modeling, analysis and control structure design. These are just a few among many applications that are likely to be pursued further in the future. It should be noted that qualitative topology has to be incorporated with quantitative information before a comprehensive application, and the topology should be adapted to different application requirements.

To develop a formal framework for these applications, we first need to formalize the description of topology; this is the objective of the next Chap. 3.

References

1. Alabi DB (2010) Automated analysis of control degree of freedom. Master's thesis, Department of chemical engineering and chemical technology, Imperial College London, London
2. Bhushan M, Rengaswamy R (2002) Comprehensive design of a sensor network for chemical plants based on various diagnosability and reliability criteria. 1. framework. Ind Eng Chem Res 41:1826–39
3. Cheng Y, Izadi I, Chen T (2013) Pattern matching of alarm flood sequences by a modified smith-waterman algorithm. Chem Eng Res Des 91(6):1085–94
4. Chiang LH, Braatz RD (2003) Process monitoring using causal map and multivariate statistics: fault detection and identification. Chemometr Intell Lab Syst 65:159–78
5. Duan P, Shah SL, Chen T, Yang F (2014) Methods for root cause diagnosis of plant-wide oscillations. AIChE J. doi:10.1002/aic.14391
6. Hangos K, Cameron I (2001) Process modelling and model analysis. Academic Press, London
7. Hangos KM, Tuza Z (2001) Optimal control structure selection for process systems. Comput Chem Eng 25(11–12):1521–1536
8. Izadi I, Shah SL, Shook D, Chen T (2009) An introduction to alarm analysis and design. In: Proceedings of 7th IFAC symposium on fault detection, supervision and safety of technical processes, Barcelona, Spain, pp 645–650
9. Jiang B, Yang F, Jiang Y, Huang D (2012) An extended AUDI algorithm for simultaneous identification of forward and backward paths in closed-loop systems. In: Proceedings of 2012 international symposium on advanced control of chemical processes, Singapore, pp 396–401.
10. Jiang B, Yang F, Huang D, Wang W (2013) Extended-AUDI method for simultaneous determination of causality and models from process data. In: Proceedings of 2013 American control conference, Washington, pp 2491–2496.
11. Jiang H, Patwardhan R, Shah SL (2009) Root cause diagnosis of plant-wide oscillations using the concept of adjacency matrix. J Process Control 19(8):1347–54
12. Nishiguchi J, Takai T (2010) Ipl2 and 3 performance improvement method for process safety using event correlation analysis. Comput Chem Eng 34(12):2007–13
13. Rothenberg D (2009) Alarm Manag Process Control. Momentem Press, New York
14. Srinivasan R, Venkatasubramanian V (1998a) Automating hazop analysis of batch chemical plants: part i. the knowledge representation framework. Comput Chem Eng 22(9):1345–5
15. Srinivasan R, Venkatasubramanian V (1998b) Automating hazop analysis of batch chemical plants: part ii. algorithms and application. Comput Chem Eng 22(9):1357–70
16. Venkatasubramanian V, Zhao J, Viswanatha S (2000) Intelligent system of hazop analysis of complex process plants. Comput Chem Eng 24:2291–302
17. Yang F, Xiao D (2012) Progress in root cause and fault propagation analysis of large scale industrial processes. J Control Sci Eng 2012:10 (Article ID 478373)
18. Yang F, Xiao D, Shah SL (2009) Optimal sensor location design for reliable fault detection in presence of false alarms. Sensors 9(11):8579–92
19. Yim SY, Ananthakumar HG, Benabbas L, Horch A, Drath R, Thornhill NF (2006) Using process topology in plant-wide control loop performance assessment. Comput Chem Eng 31(2):86–99
20. Zhang Z, Wu C, Zhang B, Xia T, Li A (2005) Sdg multiple fault diagnosis by real-time inverse inference. Reliab Eng Syst Saf 87:173–89

Chapter 3
Description of Connectivity and Causality

Abstract In this chapter, we discuss the description of two related yet different notions—connectivity and causality. Connectivity shows a physical or information linkage between process units; this linkage illustrates qualitative process knowledge without using first-principle models. The main resources for establishing connectivity are process flow diagrams (PFDs) and piping and instrumentation diagrams (P&IDs); thus we need to convert them into standard formats, such as adjacency matrices, digraphs, and semantic web models, which are easily accessible and computer-friendly. Causality between process variables can be built through process data as well as process knowledge; thus it can be described qualitatively, yet sometimes with certain quantitative information, by structural equation models, matrices and digraphs, and matrix layout plots.

Keywords Process flow diagrams · Piping and instrumentation diagrams · Adjacency matrices · Reachability matrices · Directed graphs · Semantic web · Extensible markup language · Resource description framework · Web ontology language · Structural equation models · Matrix layout plots

We begin our discussion with the description of two related yet different notions— connectivity and causality. For each of them, there are multiple formats; we will show some typical ones.

3.1 Description of Connectivity

Connectivity shows a physical or information linkage between process units; this linkage illustrates qualitative process knowledge without the needs of first-principle models. The main resource for establishing connectivity are process flow diagrams (PFDs) and piping and instrumentation diagrams (P&IDs); thus we need to convert them into standard formats that are easily accessible and computer-friendly. In what follows we introduce three main formats for this purpose.

F. Yang et al., *Capturing Connectivity and Causality in Complex Industrial Processes*, 13
SpringerBriefs in Applied Sciences and Technology,
DOI: 10.1007/978-3-319-05380-6_3, © The Author(s) 2014

3.1.1 Adjacency Matrices

An adjacency matrix [6, 7] is a matrix form to express topology with directionality. This notion of adjacency stresses that only one-step or direct connectivity is included whilst the indirect relationship is excluded because it can be inferred.

For a system with n elements $(x_k, i = 1, \ldots, n)$, an $n \times n$ adjacency matrix $\mathbf{A}$ can be defined. Each entry a_{ij} is binary: if element x_i is adjacent or directly connected to element x_j, then $a_{ij} = 1$; otherwise $a_{ij} = 0$.

Based on the adjacency matrix $\mathbf{A}$, another binary matrix, reachability matrix $\mathbf{R}$, can be derived to describe both direct and indirect relationships. Even if x_i is not adjacent to x_j, x_j may still be reached by x_i via other elements. If x_j is reached by x_i via a third element x_k, then it is called a 2-step reachability, to distinguish from the adjacency as the 1-step reachability. Similarly, k-step reachability can be defined. It can be proved that the k-step reachability can be described as the Boolean equivalent of $\mathbf{A}^k$, where the Boolean operator is defined as follows for each entry of the matrix:

$$(a_{i,j})^{\sharp} = \begin{cases} 1, & a_{i,j} \neq 0, \\ 0, & a_{i,j} = 0. \end{cases} \tag{3.1}$$

Thus, a reachability matrix is defined as:

$$\mathbf{R} = (\mathbf{A} + \mathbf{A}^2 + \cdots + \mathbf{A}^n)^{\sharp}. \tag{3.2}$$

The summation is from 1 to n because it can be proved that if two elements are not reached from one to the other via n steps, then they cannot be reached via more steps. In matrix $\mathbf{R}$, each entry r_{ij} means whether x_i can reach x_j.

Take a tank system with cascade control as an example, as shown in Fig. 3.1. To show the adjacency between each pair of elements, such as the tank, pipes, and controller, an adjacency matrix can be constructed, as shown in Fig. 3.2a. By matrix computation, one can obtain the 2-step reachability $(\mathbf{A}^2)^{\sharp}$ as shown in Fig. 3.2b, 1- or 2-step reachability $(\mathbf{A} + \mathbf{A}^2)^{\sharp}$ as shown in Fig. 3.2c, and finally the reachability matrix $\mathbf{R}$ as shown in Fig. 3.2d.

3.1.2 Digraphs

As an alternative of the adjacency matrix $\mathbf{A}$, when each element in the system is expressed by a node and each '1' entry is expressed by an arc linking two nodes corresponding to the two indices in $\mathbf{A}$, matrix $\mathbf{A}$ is converted into a directed graph or digraph including n nodes. By this conversion, the connectivity is visualized and can be better understood due to its intuitivity, because this digraph simply shows the PFD or P&ID by converting each element into an abstract node. The connectivity of the above example can be described by the digraph as shown in Fig. 3.3. Based on

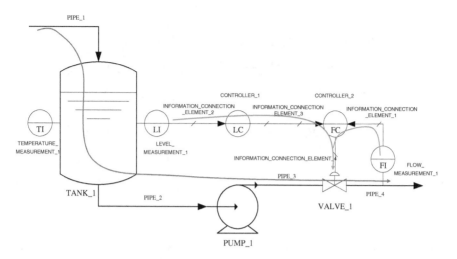

Fig. 3.1 Process schematic of a tank system with material (*blue*) and information (*red*) flow paths

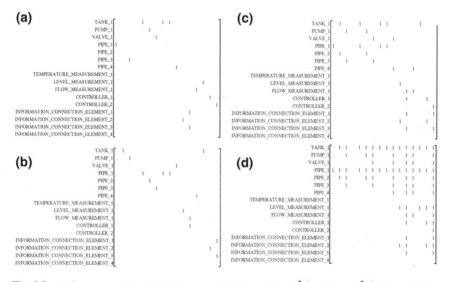

Fig. 3.2 **a** adjacency matrix **A**. **b** 2-step reachability matrix $(\mathbf{A}^2)^{\sharp}$. **c** $(\mathbf{A} + \mathbf{A}^2)^{\sharp}$. **d** reachability matrix **R**. Connectivity and reachability matrices of the tank system

this digraph, search methods in graph theory can be employed as an alternative to matrix computation, and the results can also be visualized.

To test the reachability from one node to another, a traversal search can be made to find paths between the two nodes. If there is no paths from x_i to x_j, then the corresponding entry r_{ij} in matrix **R** is '0'; otherwise, it is '1', no matter how many paths exist.

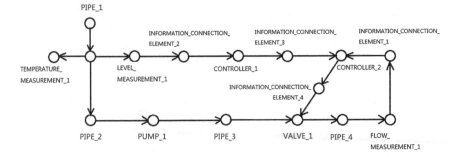

Fig. 3.3 Digraph of the tank system

The graph representation is particularly beneficial when the matrix is sparse. Moreover, some quantitative or dynamic factors can also be attached to the graph to extend its model description, which is useful in some application areas.

3.1.3 Semantic Web Description

In addition to the above mathematical descriptions, with the development of the Semantic Web, a new data model has come into use, namely, the ontology framework, which is based on a combination of artificial intelligence and database techniques. The ontology framework can be regarded as a conceptual model defined by a computer understandable language to describe and categorize the units/resources or linkages between units and their relationships. It translates the concepts defined and understood by humans into semantics in the cyber world defined by classes and rules. After this translation, new knowledge can be generated or discovered by machines through an automated inference, which makes the representation more powerful and useful [1]. By applying this technique to process modeling, the process connectivity can be modeled on the basis of PFDs or P&IDs, which facilitates the modeling and inferencing without using other special tools.

In terms of computer aided engineering exchange (CAEX) schema, eXtensible Markup Language (XML) gives users sufficient freedom to further define syntaxes and classes in their respective areas. An adjacency matrix can be constructed using the parsed information from its CAEX description—XML files [3, 9]. For the purpose of process topology description, however, a more uniform way is needed to define the process units (considered as resources) and their connections. The combination of Resource Description Framework (RDF) (http://www.w3.org/RDF/) and Web Ontology Language (OWL) (http://www.w3.org/2004/OWL/) provides a general method for conceptual description or modeling of information that is implemented in web resources, using syntaxes. In addition to connectivity, this ontological model can describe additional information such as constraints and conditions that are important for process modeling in an interoperable way.

Table 3.1 Classes of resources in the ontology framework

EQUIPMENT	UNCONTROLLED_ELEMENT	PIPE, TANK
	CONTROLLING_ELEMENT	AUTOMATIC_ELEMENT (PUMP, VALVE), MANUAL_ELEMENT (MANUAL_VALVE)
	MEASURING_ELEMENT	FLOW_MEASUREMENT, LEVEL_MEASUREMENT, TEMPERAUTRE_MEASUREMENT
COMPUTER	ANALOG_ELEMENT	CONTROLLER, SELECTOR
	DIGITAL_ELEMENT	AND, OR, NOT
	INFORMATION_CONNECTION _ELEMENT	
	AUTOMATIC_ELEMENT	PUMP, VALVE
	MEASURING_ELEMENT	FLOW_MEASUREMENT, LEVEL_MEASUREMENT, TEMPERAUTRE_MEASUREMENT

Based on the needs of process control, we first define resources by classes, which can be divided into two groups: one is equipments in the physical world, including process units and instruments; the other is computers or processors in the cyber world. Some resources can belong to both worlds, resulting in the coexistence in the two groups. From the control system perspective, sensors (transmitters), controllers, and actuators should be included in the latter category; while the sensors and actuators should also be contained in the former category because they are physical equipment. The relationship between these resources in the class domain is inheritance, namely, a subclass under a class inherits all the properties of the class; of course, a class can belong to multiple classes and inherit all the properties from them. For the tank system, a list of classes is shown in Table 3.1. Note that both the physical linkage, PIPE, and the information linkage (signal line), INFORMATION_CONNECTION_ELEMENT, are defined as classes. Next, properties are assigned to resources; these resources are the subjects of the properties. In addition to datatype and annotation properties, we define the following object properties to describe the physical and information linkages:

- uncontrolledElement.measuringElement: linkage from an uncontrolled element to a measuring element, e.g., the level of a tank measured by a sensor.
- uncontrolledElementOutlet.uncontrolledElementInlet: linkage from an uncontrolled element to another uncontrolled element, e.g., a tank connected to a pipe as an outlet.
- uncontrolledElementOutlet.controllingElementInlet: linkage from an uncontrolled element to a controlling element, e.g., a pipe connected to a control valve.
- controllingElementOutlet.uncontrolledElementInlet: linkage from a controlling element to an uncontrolled element, e.g., a valve connected to a pipe.
- computer.computer: linkage from a computer to another computer, e.g., a controller connected to a signal line (information connecting element).

The domain and range of the properties should be defined as appropriate resources.

For the tank system example (Fig. 3.1), to build the OWL file, we add instances of the above defined classes. Properties are assigned to them to define the contents and inter-relationships. For example, the outlet of PIPE_1 is connected to TANK_1; hence PIPE_1 has an object property, which is uncontrolledElementOutlet.uncontrolledElement, to have the value of another instance, TANK_1. The ontology can be visualized by OntoViz®, a plug-in for Protégé-OWL®, as shown in Fig. 3.4.

To query ontology-based RDF/OWL files, SPARQL Protocol and RDF Query Language (http://www.w3.org/TR/rdf-sparql-query/) can be used to capture useful information and conduct inferences. SPARQL uses query triples as expressions with logic operations such as conjunctions and disjunctions to perform inferences based on semantics.

One can use SPARQL to test connectivity based on object properties. If one defines a general object property and regards all the other object properties including physical and information linkages as its subproperties, then the connectivity with specified steps can be obtained. Moreover, by defining the object property as transitive, a measure of reachability can be obtained directly to show the domain of influence triggered by a change in one object.

3.2 Description of Causality

In addition to connectivity, causality between process variables should also be described. Note that the modeling resources herein include process data as well as process knowledge.

3.2.1 Structural Equation Models

Structural equation modeling (SEM) is a statistical technique for testing and estimating causal relations [8, 10]. A structural model shows potential causal dependencies between endogenous/output and exogenous/input variables, and the measurement model shows relations between latent variables and their indicators. For example, if an endogenous variable y is influenced by exogenous variables x_1 and x_2 (assume that all variables are normalized to have zero mean and unit variance), a regression model can be built as $y = p_{y1}x_1 + p_{y2}x_2 + p_{y\varepsilon}\varepsilon$ and thus be depicted as a path diagram in Fig. 3.5, where each parameter p is called a path coefficient, and ε represents the residual, that is, collective effect of all unmeasured variables that could influence y. The directed arrows represent the influence of the exogenous variables and the residual on the output variable, and the bidirectional arrow represents the correlation between exogenous variables.

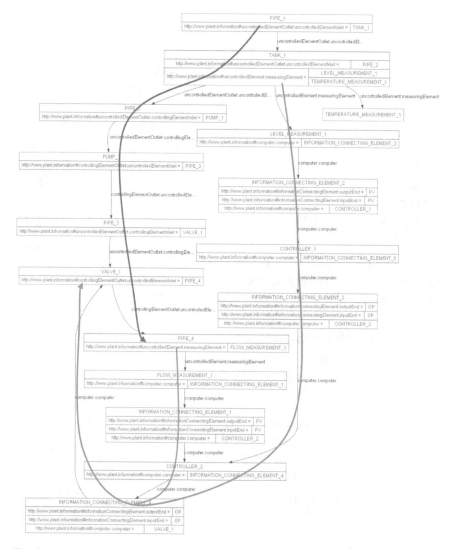

Fig. 3.4 Visualization of the tank system ontology exported by Ontoviz® with material flow (*blue*) and information flow (*red*) paths

This model is a statistical model and is highly dependant upon the partition of variables. What is more important is to obtain the topology of the system, where each variable can be both input and output variables. Thus we usually use the following descriptions.

Fig. 3.5 Path diagram of a
structural model

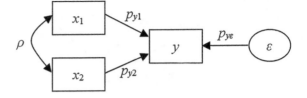

3.2.2 *Matrices and Digraphs*

Similar to the matrix or digraph formats to describe connectivity, these models can
also be used to describe causality. We have mentioned that we can introduce other
information onto the arcs of a digraph model. Typically signs can be attached to the
arcs to describe positive (promotion) or negative (inhibition) relation. For example, if
the increase (decrease) of variable x_i can cause the increase (decrease) of variable x_j,
then we define the sign as '+'. On the contrary, if the increase (decrease) of variable
x_i can cause the decrease (increase) of variable x_j, then we define the sign as '-'.
This model is called a signed digraph or signed directed graph (SDG). Normally
we use solid and broken lines to denote positive and negative relations respectively.
The formal definitions are as follows [4, 5, 11, 12]:

Definition 1: A SDG γ is the composite (G, φ) of

        (i) digraph G_0 that is the quadruple $\left(N, A, \delta^+, \delta^-\right)$ of
            (a) a set of nodes $N = \{n_1, n_2, \cdots, n_m\}$,
            (b) a set of arcs $A = \{a_1, a_2, \cdots, a_n\}$,
            (c) a couple of incident relations $\delta^+ : A \to N$ and $\delta^- : A \to N$
                that map each arc correspond to its original node and terminal
                node, respectively, and
        (ii) a function $\varphi : A \to \{+, -\}$, where $\varphi(a_k)\,(a_k \in A)$ is called the
            sign of arc a_k.

Definition 2: A pattern on the SDG model $\gamma = (G, \varphi)$ is a function $\Psi : N \to$
        $\{+, 0, -\}$. The value $\Psi(v)\,(v \in N)$ is called the sign of node v, i.e.

$$\Psi(v) = 0, \quad \text{for } |x_v - \overline{x}_v| < \varepsilon_v,$$
$$\Psi(v) = +, \quad \text{for } x_v - \overline{x}_v \geq \varepsilon_v,$$
$$\Psi(v) = -, \quad \text{for } \overline{x}_v - x_v \geq \varepsilon_v,$$

        where x_v is the measurement of the variable v, $\overline{x}_v$ is the normal value,
        and ε_v is the threshold.

Definition 3: Given a pattern Ψ on a SDG model $\gamma = (G, \varphi)$, an arc a is said to
        be consistent(with Ψ) if $\Psi\left(\delta^+a\right)\varphi(a)\Psi\left(\delta^-a\right) = +$. A path, which
        is consisted of arcs $a_1, a_2, \cdots, a_k$ linked successively, is said to be
        consistent (with Ψ) if $\Psi\left(\delta^+a\right)\varphi(a_1)\cdots\varphi(a_k)\Psi\left(\delta^-a\right) = +$.

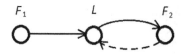

Fig. 3.6 SDG of the tank system with level control

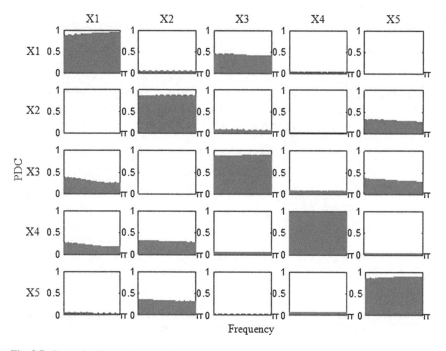

Fig. 3.7 Example of a matrix layout plot

Recall the tank system example. This time we only focus on the level control and related variables—inlet flow rate (F_1), outlet flow rate (F_2), and liquid level (L). When the level is high, the valve will open to increase the outlet flow rate according to the control law, and the result is the reduction of the level. Thus the SDG is as shown in Fig. 3.6.

The graph model is the main description of causality and we will discuss the modeling approaches and applications in the following chapters.

3.2.3 Matrix Layout Plots

Although causality is a qualitative description, it is often captured through quantitative data analysis, leading to additional information. A typical method is partial directed coherence (PDC), which has been developed and used in the neuroscience area [2]. This method can be used for multivariate systems to extract the direct causality between each pair of variables.

In the frequency domain analysis, matrix layout plots are often used, as shown in Fig. 3.7 (for details see Chap. 4). Each plot shows the information transfer from one variable to another. It is to be noted that the cause variables are listed on the top while the effect variables to be tested are on the left, which is not the same with the matrix forms mentioned earlier.

3.3 Chapter Summary

Model description is the basis of all kinds of analysis. Thus various descriptions of connectivity and causality have been briefly introduced in this chapter, which will be discussed in detail in the next chapters via different modeling approaches and other applications.

The descriptions in this chapter are limited to mathematical models and ontology models; they can be understood by computers as well as humans. The benefit is that they have potential to automate the modeling and analysis procedures. The ontology work is still ongoing, but this description has many advantages and conforms to World Wide Web Consortium (W3C) recommendations.

References

1. Allemang D, Hendler J (2011) Semantic web for the working ontologist: effective modeling in RDFS and OWL, 2nd edn. Morgan Kaufmann Publishers, San Francisco
2. Baccala LA, Sameshima K (2001) Partial directed coherence: a new concept in neural structure determination. Biol Cybern 84(6):463–474
3. Fedai M, Drath R (2005) CAEX—a neutral data exchange format for engineering data. ATP Int Autom Technol 3(1):43–51
4. Iri M, Aoki K, O'shima E, Matsuyama H (1979) An algorithm for diagnosis of system failures in the chemical process. Comput Chem Eng 3(1–4):489–493
5. Iri M, Aoki K, O'shima E, Matsuyama H (1980) A graphical approach to the problem of locating the origin of the system failure. J Oper Res Soc Jpn 23(4):295–311
6. Jiang H, Patwardhan R, Shah SL (2009) Root cause diagnosis of plant-wide oscillations using the concept of adjacency matrix. J Process Control 19(8):1347–1354
7. Mah RSH (1989) Chemical process structures and information flows. Butterworth, Boston
8. Pearl J (2009) Causality: models, reasoning, and inference, 2nd edn. Cambridge University Press, Cambridge
9. Thambirajah J, Benabbas L, Bauer M, Thornhill NF (2009) Cause-and-effect analysis in chemical processes utilizing XML, plant connectivity and quantitative process history. Comput Chem Eng 33(2):503–512
10. Wright S (1921) Correlation and causation. J Agric Res 20:557–585
11. Yang F, Xiao D (2005) Review of SDG modeling and its application. Control Theor Appl 22(5):767–774
12. Yang F, Xiao D, Shah SL (2010) Qualitative fault detection and hazard analysis based on signed directed graphs for large-scale complex systems. In: Zhang W (ed) Fault Detection. In-Tech, Vukovar, Crotia, pp 15–50

Chapter 4
Capturing Connectivity and Causality from Process Knowledge

Abstract Process knowledge is the most reliable resource for qualitative modeling of complex industrial processes, which is typically expressed in natural language and stored in human brains. We thus need to capture useful connectivity and causality from such resources and convert the information into computer accessible formats. From first-principle structural models, causality can be captured and expressed as structural equations. From unstructured process knowledge and dynamic and algebraic equations, graphical models, in particular signed directed graphs and variants, can be obtained. Graphic models are widely used due to their computer tractability and human readability. Rule-based models are another alternative, which is used in expert systems. When the process information is accessible in web language, connectivity can be retrieved by query.

Keywords Process knowledge · First-principle models · Structural equation modeling · Signed directed graphs · Dynamic and algebraic equations · Perfect matching · Control laws · Bond graphs · Expert systems

Process knowledge is the most reliable resource for qualitative modeling of complex industrial processes, which includes PFDs, P&IDs, and other expert knowledge that is typically expressed in natural language and stored in human brains. In many cases, process knowledge is not limited to qualitative information as first-principle models or even grey-box models may also be available. Either qualitative information or quantitative description is valuable resource to capture qualitative connectivity and causality, even though some of the information may be missing. Our task is to capture useful connectivity and causality from such resources and convert the information into computer accessible formats.

F. Yang et al., *Capturing Connectivity and Causality in Complex Industrial Processes*, 23
SpringerBriefs in Applied Sciences and Technology,
DOI: 10.1007/978-3-319-05380-6_4, © The Author(s) 2014

4.1 Structural Modeling Based on First-Principle Structural Models

A structural model shows potential causal dependencies between endogenous/output and exogenous/input variables, and the measurement model shows relations between latent variables and their indicators. As mentioned in Sect. 3.2.1, a directed arrow represents the influence of an exogenous variable or the residual on the output variable, and a bidirectional arrow represents the correlation between exogenous variables.

Since the exogenous variables are not independent, there is some ambiguity about the real or dominant path. Structural equation modeling (SEM) is a statistical technique for testing and estimating causal relations [22, 25]. Based on the statistical analysis, components of direct and indirect relations can be evaluated via variance decomposition [31]; this provides some indication of the model structure. Typically, factor analysis, path analysis, and regression, as special cases of SEM, are widely used in exploratory factor analysis, such as psychometric design. IBM® SPSS® Amos (Analysis of Moment Structures) provides an easy-to-use program for visual SEM.

The limitations of this modeling approach are: (1) exogenous and endogenous variables should be selected in advance as a hypothesis and the result highly depends on this partition; (2) the causal relations described here are static relations; and (3) only linear regression is considered. To overcome the last two limitations, dynamic causal modeling embraces nonlinear and dynamic relationships [8]. This approach is more suitable for confirmatory modeling than exploratory modeling to construct a network topology, and suffers from a high dimension problem involving a large number of variables; for this reason, we include it in this category although it is essentially a combination of first principles and process data.

4.2 Construction of Adjacency and Reachability Matrices

As defined in Sect. 3.1.1, an adjacency matrix can be built manually based on process connectivity knowledge. If the PFD or P&ID diagram is available, this is straightforward. By Boolean matrix computation, a reachability matrix can be obtained. This can also be considered as an equivalent representation of a process graph [13].

4.3 Construction of Graphical Models

4.3.1 Modeling of SDGs

As an extension of process graphs, SDGs are established by representing process variables as graph nodes and representing causal relations as directed arcs. An arc from node x to node y implies that the variation of x may cause the variation of y. For convenience, '+', '−', or '0' is assigned to the nodes in comparison with normal operating value thresholds to denote higher than, lower than, or within the normal region, respectively. Positive or negative influence between nodes is distinguished by the sign '+' (promotion) or '−' (suppression), assigned to the arc.

A SDG can be built manually from first principles and mathematical models and more practically from process knowledge including flowsheets [14, 15]. If we have the differential algebraic equations (DAEs) as the process description, then we can derive the structure and signs of the graph by specific methods briefly introduced below [14].

A typical dynamic system can be expressed as a set of differential equations (DEs)

$$\frac{dx_i}{dt} = f_i(x_1, \ldots, x_n), \quad i = 1, \ldots, n, \tag{4.1}$$

where $x_1, \ldots, x_n$ are process variables. By Taylor series expansion near an operating point $x_1^0, \cdots, x_n^0$, we obtain

$$\frac{dx_i}{dt} \approx f_i\left(x_1^0, \ldots, x_n^0\right) + \sum_{j=1}^n \left.\frac{\partial f_i}{\partial x_j}\right|_{x_1^0, \ldots, x_n^0} \left(x_j - x_j^0\right), \tag{4.2}$$

where $f_i(x_1^0, \ldots, x_n^0) = 0$. Equation (4.2) can be expressed in the following matrix form:

$$\frac{d}{dt}\begin{bmatrix} x_1 \\ \vdots \\ x_n \end{bmatrix} \approx \begin{bmatrix} \frac{\partial f_1}{\partial x_1} & \cdots & \frac{\partial f_1}{\partial x_n} \\ \vdots & & \vdots \\ \frac{\partial f_n}{\partial x_1} & \cdots & \frac{\partial f_n}{\partial x_n} \end{bmatrix}\Bigg|_{x_1^0, \ldots, x_n^0} \begin{bmatrix} x_1 - x_1^0 \\ \vdots \\ x_n - x_n^0 \end{bmatrix}. \tag{4.3}$$

The Jacobian matrix

$$J = \begin{bmatrix} \frac{\partial f_1}{\partial x_1} & \cdots & \frac{\partial f_1}{\partial x_n} \\ \vdots & & \vdots \\ \frac{\partial f_n}{\partial x_1} & \cdots & \frac{\partial f_n}{\partial x_n} \end{bmatrix} \tag{4.4}$$

can be described by a SDG whose signs of arcs are defined as

$$\text{sgn}(x_j \rightarrow x_i) = \text{sgn}\left[\left.\frac{\partial f_i}{\partial x_j}\right|_{x_i^0, \ldots, x_n^0}\right], \tag{4.5}$$

if the nodes correspond to the process variables. Thus the SDG in fact describes the direct influences or sensitivities between process variables.

Fig. 4.1 Step response of different systems. **a** First-order system, **b** Higher-order system with its approximate first-order system

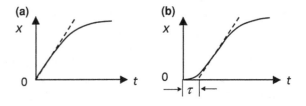

In practical problems, the systems often have the following form as a DE:

$$a_n \frac{d^n x}{dt^n} + \cdots + a_2 \frac{d^2 x}{dt^2} + a_1 \frac{dx}{dt} + a_0 x = e, \tag{4.6}$$

where x is the state and e is the disturbance. When $n = 1$, it is a first-order system:

$$\frac{dx}{dt} = -\frac{a_0}{a_1} x + \frac{1}{a_1} e. \tag{4.7}$$

The step response is shown in Fig. 4.1a. An arc is constructed from node e to node x with a sign sgn$[1/a_1]$ and a self-cycle on the node x with a sign -sgn$[a_0/a_1]$. For higher-order systems, simplification can be made because the corresponding DE includes different order derivatives of the same variable, which can be avoided for the explicit physical meaning of the model. It can be approximated as a first-order system with delay:

$$\frac{d}{dt} x(t - \tau) = -\frac{a_0'}{a_1'} x(t) + \frac{1}{a_1'} e(t), \tag{4.8}$$

where τ is the equivalent pure delay. Its step response is shown in Fig. 4.1b. The method of constructing SDGs is the same as the former one, and the delay can be included in dynamic SDGs [27].

Algebraic equations (AEs) are usually included in the mathematical models as constraints and can also be transformed into SDGs [14] although they are non causal in nature. Because there may exist multiple perfect matchings between equations and variables, the corresponding SDGs may not be unique. Some treatment should be made to remove the unsteady or spurious SDGs [14, 19].

Take a two-tank system as an example, as shown in Fig. 4.2a. Two tanks are connected by a pipe; both tanks have outlet pipes, and Tank 1 has a feed flow. This system can be described by the following set of DAEs:

$$C_1 \frac{de_2}{dt} = f_1 - f_3 - f_5,$$

$$C_2 \frac{de_7}{dt} = f_5 - f_8,$$

$$f_3 = \frac{1}{R_{b1}} \sqrt{l_2}, \tag{4.9}$$

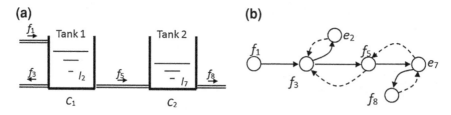

Fig. 4.2 Two-tank system. **a** Schematic. C_1 and C_2 are cross-sectional areas of the two tanks respectively. **b** SDG. e_2 and e_7 are the square roots of levels in the two tanks respectively

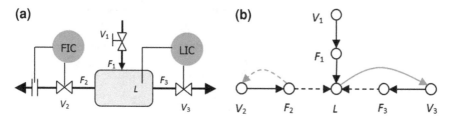

Fig. 4.3 Schematic and SDG of a tank system with controlled flowrates. **a** Schematic, **b** SDG

$$f_5 = \frac{1}{R_{12}}(\sqrt{l_2} - \sqrt{l_7}),$$

$$f_8 = \frac{1}{R_{b2}}\sqrt{l_7},$$

where l_2 and l_7 are the levels in Tanks 1 and 2, f_1, f_3, f_5, and f_8 are flowrates, and R_{12}, R_{b1}, and R_{b2} are the pipe resistances between Tanks 1 and 2 and the two outlet pipes, respectively. Since l_i ($i = 2$ or 7) appears as the square root form, we use e_i to denote the square root. One can convert these equations to nodes and arcs to form a SDG, as shown in Fig. 4.2b, where solid lines denote positive influences and broken lines denote negative influences. Although no control is taken, there are still some signal paths as shown in Fig. 4.2b.

In many cases, the SDGs are established by qualitative process knowledge and experience. Figure 4.3a shows a tank with one inlet and two outlets, both under control. The arcs from F_2 to V_2 and L to V_3 in Fig. 4.3b describe the flow control and level control loops respectively. Each control loop can be expressed by a negative cycle in SDG because of the negative feedback action. This qualitative SDG can be obtained directly from process knowledge and does not need the exact mathematical equations. Sometimes the qualitative simulation and sensitivity experiments may also help. The SDGs obtained by this method often include indirect causalities in addition to direct ones, so the graph should be simplified and transformed so that all the arcs stand for direct causalities. Some rules are summarized in [26].

The modeling procedure is often step by step. Take a continuous stirred tank reactor (CSTR) system as another example, as shown in Fig. 4.4.

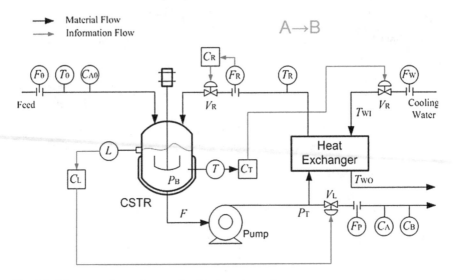

Fig. 4.4 Schematic of a continuous stirred tank reactor system

First, the level in the tank is influenced by the inlet, outlet and recycle, so these four variables have the relationship similar to the above examples (Fig. 4.5a). Next, the outflow is associated with the reactor inner pressure P_B and outlet pipe pressure P_T (Fig. 4.5b). For the heat exchanger, the temperatures are influenced by flowrates (Fig. 4.5c). Since the reaction is irreversible and exothermic, the relations with concentrations should be included (Fig. 4.5d). When measurement and control are taken into account, the complete SDG is shown in Fig. 4.5e.

The SDG model can be validated by process data [29]. For example, correlation is a necessary condition of causality, so the cross-correlation between every two measured variables can be used to validate the arcs in SDGs, and the directions can also be obtained by shifting the time series (adding lags) to find the maximal cross-correlation. Alternatively, probabilistic measure, such as transfer entropy, can be used to obtain the causality and directionality [2]. These data-based methods will be discussed in the next chapter.

In summary, the main steps of SDG modeling are: (1) Collect process knowledge, especially P&IDs and equations; (2) build the material flow digraph by connectivity information between entities; (3) choose the key variables and give them signs according to process knowledge; (4) add control arcs on the digraph to constitute the SDG skeleton; (5) add other variables and arcs to form the entire SDG; (6) simplify and verify the SDG by graph theory; and, finally, (7) validate the SDG with process data and sensitivity experiments.

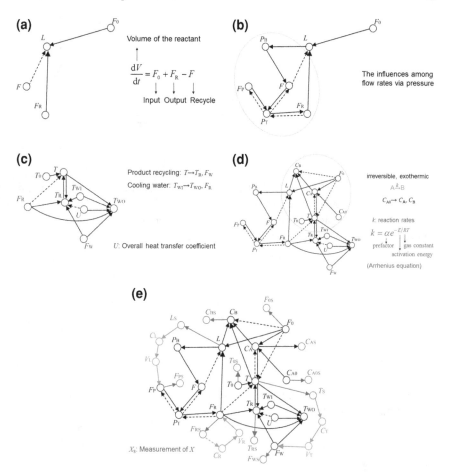

Fig. 4.5 SDG modeling procedure of the CSTR system. **a** Unit SDG regarding the tank, **b** extended SDG considering the pressure, **c** unit SDG regarding the heat exchanger, **d** extended SDG considering the concentrations, **e** extended SDG considering measurement and control

4.3.2 SDG Modeling in Control Systems

Control systems are a necessary part in industrial processes. The SDG modeling of common controls will be discussed below by using the above method [28].

In control applications, PID control is the most common mode of control. As shown in Fig. 4.6, a control loop is composed of a sensor, a controller, an actuator, and a controlled plant. The deviation e of the set point r and the measurement value x_m of the controlled variable x, is the input signal to the controller, and the output of the controller u is sent to the actuator and thus affects the controlled plant through the manipulative variable q. This constitutes a closed loop. Because the controlled

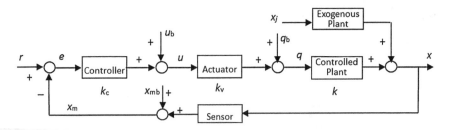

Fig. 4.6 Block diagram of a feedback control loop

variable may be affected by some disturbances or be coupled with other system variables, the exogenous plant and variable x_i are also added in Fig. 4.6. Assume that the controlled plant and the controller are both linear amplifiers, namely, proportion elements, with the positive gains k and k_y, respectively. The control law of a PID controller is:

$$\begin{cases} u = u_P + u_I + u_D, \\ u_P = k_c e, \\ \dfrac{du_I}{dt} = \dfrac{k_c e}{\tau}, \\ u_D = k_c \tau_D \cdot \dfrac{de}{dt}, \end{cases} \qquad (4.10)$$

where, k_p is the positive proportion parameter, τ_I and τ_D are integral and differential time constants, respectively.

According to the control law, the DAEs of the system are as follows:

$$x_m = x + x_{mb}, \qquad (4.11)$$

$$e = r - x_m, \qquad (4.12)$$

$$u_P = k_c e, \qquad (4.13)$$

$$\frac{du_I}{dt} = \frac{k_c e}{\tau_I}, \qquad (4.14)$$

$$u_D = k_c \tau_D \cdot \frac{de}{dt}, \qquad (4.15)$$

$$u = u_P + u_I + u_D + u_b, \qquad (4.16)$$

$$q = k_v u + q_b, \qquad (4.17)$$

$$x = k_q + a_j x_j, \qquad (4.18)$$

where subscript 'b' denotes bias. In the AE portion, there are two "perfect matchings" [14] between the set of equations and the set of variables, as shown in Table 4.1, whose corresponding SDGs are shown in Fig. 4.7, in which the nodes with shadow are deviation nodes, arrows with solid and dotted lines denote signs '+' and '−', respectively. It is noted that the node de/dt is an individual node with special function,

Table 4.1 Perfect matchings between the AEs and variables

Equations	Matched variables in perfect matching (case 1)	Matched variables in perfect matching (case 2)
(4.11)	x_m	x
(4.12)	e	x_m
(4.13)	u_P	e
(4.15)	u_D	u_D
(4.16)	u	u_P
(4.17)	q	u
(4.18)	x	q

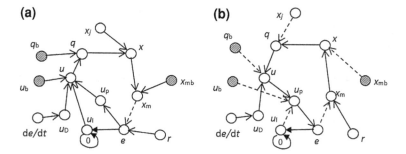

Fig. 4.7 Two SDGs of the PID control loop. **a** Case 1, and **b** case 2

although it is the derivative of e. In applications, we generally assume that all changes on nodes are step functions, because the SDGs are only used to analyze qualitative trends. Hence de/dt can also be replaced by e, but its effect is limited to initial response. Here the effect of de/dt on u_D is the same as the effect of e on u_P, but with a shorter duration.

Equation (4.13) describes the controlled plant; thus the arc direction should be from q to x according to the physical meaning, which shows the cause-effect relationship; so the case of Fig. 4.7b is removed. Moreover, if the plant shows some dynamic characteristic, for example, the left-hand of the equation is dx/dt, then the equation becomes a DE, and hence there is only one perfect matching, and the case of Fig. 4.7b does not exist anymore. Using Fig. 4.7a, the initial response can be analyzed, for example, if the set point r is increased, then e, u_P, u, q, x, and x_m will become '+' immediately, and u_I will become '+' gradually because the arc from e to u_I is a DE arc. This propagation path $r \rightarrow e \rightarrow u_P \rightarrow u \rightarrow q \rightarrow x \rightarrow x_m$ is consistent with the actual relations of information transfer. No matter whether the case of Fig. 4.7b is reasonable, the analysis results of initial response by the two SDGs are the same because there are no positive cycles within them. We summarize this as the following rule:

Fig. 4.8 Steady-state SDG of
a PID control loop

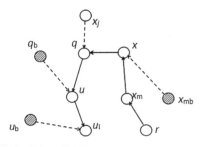

Rule 1: The fault propagation path of the initial response in a control loop is the longest acyclic path starting from the fault origin in the path "set point $\rightarrow$ error $\rightarrow$ controlling variable $\rightarrow$ controlled variable $\rightarrow$ measurement value $\rightarrow$ error", which is consistent with the information flow in the block diagram.

For the final response, the left-hand side of (4.14) is zero, so $e = 0$ in the steady state, which can be obtained from the concept. Hence u_P and u_D are both zeros. The above DAEs can be transformed into:

$$x_m = x + x_{mb}, \tag{4.19}$$

$$x_m = r, \tag{4.20}$$

$$u = u_I + u_b, \tag{4.21}$$

$$q = k_v u + q_b, \tag{4.22}$$

$$x = kq + a_j x_j. \tag{4.23}$$

Now the perfect matching is exclusive and the corresponding SDG is shown in Fig. 4.8 that is the simplification of Fig. 4.7b. There are two fault propagation paths: $r \rightarrow x_m \rightarrow x \rightarrow q$ and $x_j \rightarrow q \rightarrow u \rightarrow u_I$. If the set point r is increased, then x_m, x, q, u, and u_I will all be increased in the steady state as long as the control action is effective. However, if only x_{mb} is increased, then x_m will not be affected, but x will be increased, that is the action of the control loop. We find that Fig. 4.7b also makes sense as it reflects the information flow in steady state. From a physical viewpoint, when the control loop operates, the controlled variable is determined by the set point, and the controller looks like an amplifier with an infinite gain, whose input equals zero and whose output is determined by the set point. This logical transfer relation is opposite to the actual information relation.

Because the D action is restricted in the initial period, the fault propagation path of PI control is the same as the one above. Because of I action, some variables show compensatory response, for example, the response of x_m due to x_{mb} is limited at initial stage. If there is only P action, then e is not zero in the steady state, and thus u_I and related arcs in Fig. 4.8 are deleted, and the initial response and steady-state response can both be analyzed with this graph.

The rule of fault propagation analysis in steady state can be summarized as follows:

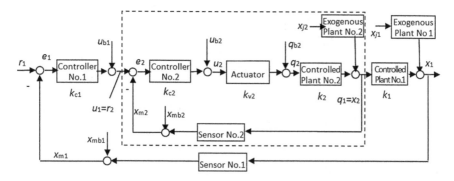

Fig. 4.9 Block diagram of a cascade control system

Rule 2: The fault propagation path of the steady-state response in a control loop is the path "set point $\rightarrow$ measurement value $\rightarrow$ controlled variable $\rightarrow$ controlling variable" and "exogenous variable $\rightarrow$ controlling variable".

When the control loop works, the above analysis shows the fault propagation paths due to the output deviation of the sensor, controller, actuator, and other exogenous variables. When the control loop does not function, there are two cases: (1) structural faults, e.g., the failure of the sensor, controller or actuator causes a broken arc and the control loop becomes open, (2) excessive deviation causes the controller saturation, leading to the I action ineffective in removing the residual error, which is similar to the P action case.

Based on the above analysis of the PID control loop, other control systems can also be transformed into SDG forms.

Feedforward control is a supplement to feedback control. It is widely used in practice; it is easy to be treated according to the foregoing methods because it composes paths but not cycles, hence leading to no multiple perfect matchings.

Split-range control means different control strategies are adopted at different value intervals. Here the sign of the arcs or even the graph structure may change with the variable values, which is realized by several controllers in a parallel connection. This case is hard for SDG to deal with. We have to make some judgments as adding inference, and modify the structure or use conditional arcs to cover all the cases [23].

Cascade control can be regarded as an extension of the single loop case. It can be implemented directly by AEs, or by the combination of two single loops. For example, the cascade control system in Fig. 4.9 has the steady-state SDG as shown in Fig. 4.10, where the controlled variable of the outer loop u_1 is the set point of the inner loop r_2.

Other control methods include ratio control and averaging control. Figure 4.11 is a dual-element averaging control system whose objective is to balance two variables—level and flow, the block diagram of which is shown in Fig. 4.12. We have

$$P_x = P_L - P_F + P_S + c, \tag{4.24}$$

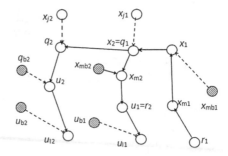

Fig. 4.10 Steady-state SDG of a cascade control system

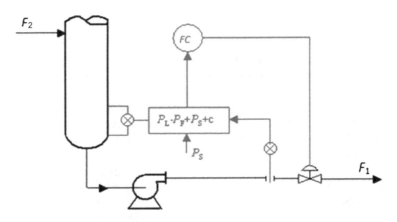

Fig. 4.11 Dual-element averaging control system

Fig. 4.12 Block diagram of a dual-element averaging control system

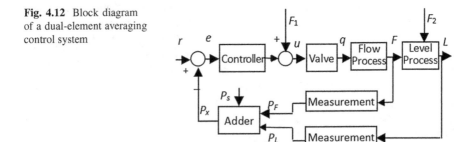

where P_x is the pressure signal of the adder output, P_L is the level measurement signal, P_F is the flow measurement signal, and P_S is a tunable signal of the adder. In the simplest case, flow process and its measurement are both positive linear elements, and the level process is a negative linear element, so the steady-state SDG is shown

Fig. 4.13 Steady-state SDG
of a dual-element averaging
control system

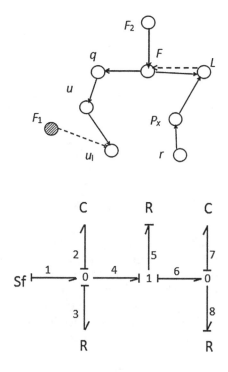

Fig. 4.14 Bond graph of a
two-tank system

in Fig. 4.13. Although there are multiple perfect matchings, the SDG has only one
negative cycle; thus we can analyze the fault propagation through the directed paths.

Thus we conclude the following rule:

Rule 3: The fault propagation path of a control system in steady state can be
combined from the ones of single-loop by combining the same nodes and adding
arcs by transforming AEs that describe the relationships.

4.3.3 Other Graphical Models

There are other graphical models that are commonly used to describe complex systems, and yet they have different forms with different meanings. Bond graphs [21]
and their extensions, such as temporal causal graphs [18], use different symbols to
further describe dynamic characteristics. More precisely, qualitative transfer functions [12], differential equations [17], and trend analysis [9, 16] have been integrated
into causal graphs, and complex algorithms are introduced to improve their correct-
ness [4]. Similar or improved approaches have been investigated by many researchers
[1, 6, 10, 20].

The bond graph of the two-tank system is shown in Fig. 4.14, and the temporal causal graph is shown in Fig. 4.15. In the bond graph, there are two types of

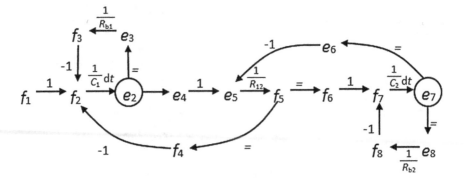

Fig. 4.15 Temporal causal graph of the two-tank system

junctions—common effort (0-) junctions and common flow (1-) junctions. The bond
graph describes the exchange of physical energy by bonds. A bond graph can be used
to derive the steady state model automatically; this property is similar with signal
flow graphs, which, as another graphical model, can be used for derivation of transfer
functions. The temporal causal graph converts the junctions and bonds in the bond
graph into nodes and arcs, and imposes labels on arcs to describe detailed temporal
effects such as integration and rate of change.

Compared to the SDG in Fig. 4.2b, the temporal graph provides more detailed
information and forms a quantitative model, while the SDG is only concerned with
the qualitative trends. Since exact models are often difficult to obtain for industrial
processes, SDG models are more appealing due to their simplicity.

4.4 Rule-Based Models

In addition to the above models to describe connectivity and causality information
from process knowledge, there are other methods that can be used, such as rules in
expert systems.

An expert system is a powerful tool in artificial intelligence by converting knowledge into a set of IF-THEN rules. Kramer and Palowitch [11] used such rules to
describe SDG arcs and thus expert systems can be employed as a tool in this problem. Each arc can be described by a rule using logical functions p, m, and z:

$$(pAB) \Leftrightarrow A \longrightarrow B \text{ (positive relation)},$$
$$(nAB) \Leftrightarrow A \dashrightarrow B \text{ (negative relation)}, \qquad (4.25)$$
$$(zAB) \Leftrightarrow A \qquad B \text{ (zero relation)}.$$

Therefore, a SDG can be converted into a set of rules. These rules can be expressed
in IF-THEN forms to make reasoning by rule reduction. Since only qualitative

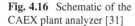

Fig. 4.16 Schematic of the CAEX plant analyzer [31]

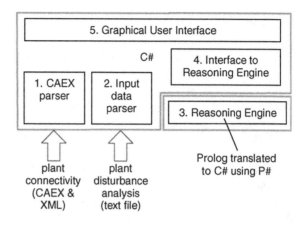

information is included, there may exist a lot of illusive results. To prevent this disadvantage, some quantitative information, such as steady-state gain, is taken into account to find dominant propagation paths [3].

4.5 Extracting Plant Topology from Web Language

P&IDs and other flowsheets are very important topological process knowledge expression that can be standardized in a computer readable electronic form—XML format. It has been implemented in some commercial CAD tools such as Aveva's VPE P&ID, Comos P&ID from Siemens, and SmartPlant P&ID from Intergraph according to standards IEC/PAS 62424, ISO 10303-221 and ISO 15926 on the exchange of engineering data [7]. The XML-based CAEX provides a common data format for storage of object information. A tool has been developed to parse and extract the connectivity information from an XML file [24, 31], named as the CAEX Plant Analyzer. The schematic of this tool is shown in Fig. 4.16, where the CAEX & XML can be parsed as well as plant disturbance analysis results for reasoning.

In Sect. 3.1.3, the RDF/OWL techniques were introduced as a semantic web description of plant topology. This format includes more information and can be used to query for more information. The topology information is a simple application and can be extracted easily. With the progress of semantic web technology, this emerging technique will have more applications.

The topology or connectivity obtained here includes both material flow paths and information flow paths, which are needed for topology modeling. Although the granularity is entity-based, which is not enough for the variable-based topology modeling, this kind of topological information is fundamental and can be used as references as well.

4.6 Chapter Summary

Some common approaches to topology capture from process knowledge have been introduced in this chapter, particularly the SDG model building. This procedure is tedious and time-consuming because it is usually a manual procedure. Web language extracted from CAD tools provides an opportunity to automate this procedure and some preliminary work has shown this feasibility.

In practice, a large-scale system is composed of many elementary units, such as tanks and heat exchangers. These generic unit models from various unit operations can be developed and stored in a library for reuse. When building a model for a new process, one can just connect the individual models using the connectivity information from the plant topology by matching outlet variables of the upstream elements to the input variables of the downstream elements [5]. In addition, hierarchical models can be developed using the concept of divide and conquer [30].

All these approaches need the complete knowledge of the process, at least the qualitative knowledge, which is not always available; thus we should also explore topology capture using data-based methods.

References

1. Alonso CJ, Llamas C, Maestro JA, Pulido B (2003) Diagnosis of dynamic systems: a knowledge model that allows tracking the system during the diagnosis process. Lect Notes Artif Intell 2718(6):208–218
2. Bauer M, Cox JW, Caveness MH, Downs JJ, Thornhill NF (2007) Finding the direction of disturbance propagation in a chemical process using transfer entropy. IEEE Trans Control Syst Technol 15(1):12–21
3. Chang CC, Yu CC (1990) On-line fault diagnosis using the signed directed graph. Ind Eng Chem Res 29(7):1290–1299
4. Cheng H, Tikkala VM, Zakharov A, Myller T, Jamsa-Jounela SL (2011) Application of the enhanced dynamic causal digraph method on a three-layer board machine. IEEE Trans Control Syst Technol 19(3):644–655
5. Di Geronimo Gil GJ, Alabi DB, Iyun OE, Thornhill NF (2011) Merging process models and plant topology. In: Proceedings of 4th advanced control of industrial processes, Hangzhou, China
6. Fagarasan I, Ploix S, Gentil S (2004) Causal fault detection and isolation based on a set-membership approach. Automatica 40(12):2099–2110
7. Fedai M, Drath R (2005) CAEX—a neutral data exchange format for engineering data. ATP Int Autom Technol 3(1):43–51
8. Friston KJ, Harrison L, Penny W (2003) Dynamic causal modelling. NeuroImage 19(4):1273–1302
9. Gao D, Wu C, Zhang B, Ma X (2010) Signed directed graph and qualitative trend analysis based fault diagnosis in chemical industry. Chinese J Chem Eng 18(2):265–276
10. Jan A, Jonas B, Erik F, Krysander M, Lars N (2007) Safety analysis of autonomous systems by extended fault tree analysis. Int J Adapt Control Signal Process 21(2–3):287–298
11. Kramer MA, Palowitch BL Jr (1987) A rule-based approach to fault diagnosis using the signed directed graph. AIChE J 33(7):1067–1078
12. Leyval L, Gentil S, Feray-Beaumont S (1994) Model based causal reasoning for process supervision. Automatica 30(8):1295–1306

13. Mah RSH (1989) Chemical process structures and information flows. Butterworth, Boston, MA
14. Maurya MR, Rengaswamy R, Venkatasubramanian V (2003) A systematic framework for the development and analysis of signed digraphs for chemical processes. 1. Algorithms and analysis. Ind Eng Chem Res 42(20):4789–4810
15. Maurya MR, Rengaswamy R, Venkatasubramanian V (2003) A systematic framework for the development and analysis of signed digraphs for chemical processes. 2. Control loops and flowsheet analysis. Ind Eng Chem Res 42(20):4811–4827
16. Maurya MR, Rengaswamy R, Venkatasubramanian V (2007) A signed directed graph and qualitative trend analysis-based framework for incipient fault. Chem Eng Res Des 85(10):1407–1422
17. Montmain J, Gentil S (2000) Dynamic causal model diagnostic reasoning for online technical process supervision. Automatica 36(8):1137–1152
18. Mosterman PJ, Biswas G (1999) Diagnosis of continuous valued systems in transient operating regions. EEE Trans Syst Man Cybern Part A 29(6):554–565
19. Oyeleye OO, Kramer MA (1988) Qualitative simulation of chemical process systems: steady-state analysis. AIChE J 34(9):1441–1454
20. Pastor J, Lafon M, Trave-Massuyes L, Demonet JF, Doyon B, Celsis P (2000) Information processing in large-scale cerebral networks: the causal connectivity approach. Biol Cybern 82(1):49–59
21. Paynter HM (1960) Analysis and design of engineering systems. MIT Press, Cambridge, MA
22. Pearl J (2009) Causality: models, reasoning, and inference, 2nd edn. Cambridge University Press, Cambridge, UK
23. Shiozaki J, Matsuyama H, O'Shima E, Iri M (1985) An improved algorithm for diagnosis of system failures in the chemical process. Comput Chem Eng 9(3):285–293
24. Thambirajah J, Benabbas L, Bauer M, Thornhill NF (2009) Cause-and-effect analysis in chemical processes utilizing XML, plant connectivity and quantitative process history. Comput Chem Eng 33(2):503–512
25. Wright S (1921) Correlation and causation. J Agric Res 20:557–585
26. Yang F, Xiao D (2005) Approach to modeling of qualitative SDG model in large-scale complex systems. Control Instrum Chem Ind 32(5):8–11
27. Yang F, Xiao D (2006) Approach to fault diagnosis using SDG based on fault revealing time. Proceedings of 6th world congress on intelligent control and automation, Dalian, China, pp 5744–5747
28. Yang F, Shah SL, Xiao D (2009) SDG model-based analysis of fault propagation in control systems. Proceedings of 2009 Canadian conference on electrical and computer engineering, St John's, NL, Canada, pp 1152–1157
29. Yang F, Shah SL, Xiao D (2012) Signed directed graph based modeling and its validation from process knowledge and process data. Int J Appl Math Comput Sci 22(1):41–53
30. Yang F, Xiao D, Shah SL (2013) Signed directed graph-based hierarchical modelling and fault propagation analysis for large-scale systems. IET Control Theory Appl 7(4):537–550
31. Yim SY, Ananthakumar HG, Benabbas L, Horch A, Drath R, Thornhill NF (2006) Using process topology in plant-wide control loop performance assessment. Comput Chem Eng 31(2):86–99

Chapter 5
Capturing Causality from Process Data

Abstract Data is a valuable resource for modeling and analysis. Process data is a set of timeseries of process variables. In this chapter, we focus on the relationship between different time series to capture causality in the process. For a pair of process variables, various data-based methods can be applied to detect causality. These methods can be categorized into three classes: lag-basedmethods, such as the Granger causality and transfer entropy; conditional independence methods, such as the Bayesian network; and higher order statistics, such as the Patel's pairwise conditional probability approach. In this work, we focus on the first group of methods, which are the most commonly used, and then briefly discuss some remaining methods. Based on the results of pairwise causality analysis, one can construct a causal network that is composed of the links between every two nodes. For multivariate systems, network topology can be determined by using statistical confounding analysis.

Keywords Big data · System identification · Cross correlation · Correlation color maps · Granger causality · Directed transfer functions · Partial directed coherence · Transfer entropy · Bayesian network · Mutual validation

Data is a valuable resource for modeling and analysis. Process data is a set of time series of process variables. In this chapter, we focus on the relationship between different time series to capture causality in the process. Although the concept of big data concentrates on correlation instead of causality [23], we still need to detect and infer causality in the analysis of industrial processes with clear physical background.

In Chap. 1, we explained the notion of causality. For a pair of process variables, various data-based methods can be applied to detect causality. These will be introduced in the following sections. However, there are a lot of process variables in a process. Based on the results of pairwise causality analysis, one can construct a causal network that is composed of the links between every two nodes. If there are n nodes in this network, there are n^2 links that need to be checked. Because causality is transitive, some causality relations can be explained by sequential direct causal relations. For example, the causality from A to C can be a combined result of causal relations from A to B and from B to C; pairwise data analysis cannot recognize this difference. For such cases network topology can be determined by using statistical confounding analysis.

These methods can be categorized into three classes [34]: lag-based methods, such as Granger causality and transfer entropy; conditional independence methods, such as Bayesian network; and higher order statistics, such as Patel's pairwise conditional probability approach [27]. In this work, the emphasis will be on the first group of methods, which is the most commonly used, and briefly discuss some remaining methods.

5.1 System Identification Approach

If we consider a pair of variables as a bivariate system, the closed-loop system identification methods can be applied to determine the path models. By checking the existence of the paths, we can conclude the causality between the two variables. Under the assumption of the model structure, both the model orders and parameters need to be identified; to do this we propose the use of the augmented upper diagonal identification (AUDI) algorithm [24] and its extension—the interleaved data pair upper diagonal (IDPUD) algorithm [17]—to identify the model order and the parameters simultaneously.

Consider the following linear closed-loop model

$$
\begin{aligned}
A(z^{-1})z(k) &= B(z^{-1})u(k) + v(k), \\
P(z^{-1})u(k) &= Q(z^{-1})z(k) + w(k),
\end{aligned}
\tag{5.1}
$$

with

$$
\begin{cases}
A(z^{-1}) = 1 + a(1)z^{-1} + \cdots + a(n_a)z^{-n_a}, \\
B(z^{-1}) = b(d)z^{-d} + \cdots + b(n_b)z^{-n_b}, \\
P(z^{-1}) = p(1)z^{-1} + \cdots + p(n_p)z^{-n_p}, \\
Q(z^{-1}) = q(c)z^{-c} + \cdots + q(n_q)z^{-n_q},
\end{cases}
\tag{5.2}
$$

where $z(k)$ and $u(k)$ are the output and input variables; $v(k)$ and $w(k)$ are white noise sequences; $a(i)$ and $b(i)$ are parameters of the forward path model; $p(i)$ and $q(i)$ are parameters of the backward path model; n_a, n_b, n_p, and n_q are the corresponding orders; c and d are the delays in the backward and forward paths, respectively.

Define the following interleaved data vectors as

$$
\begin{cases}
\varphi(k) = [-z(k-n), u(k-n), \ldots, -z(k-1), -u(k-1), -z(k)]^T, \\
\mathbf{h}(k) = [-z(k-n), u(k-n), \ldots, -z(k-1), -u(k-1)]^T, \\
\mathbf{g}(k) = [-u(k-n), z(k-n), \ldots, -u(k-1), -z(k-1)]^T,
\end{cases}
\tag{5.3}
$$

where

$$n \geqslant \max \{n_a, n_b, n_p, n_q\}. \tag{5.4}$$

Then (5.1) can be rewritten as

$$z(k) = \mathbf{h}^T(k) \cdot \theta + v(k), \tag{5.5}$$

$$u(k) = \mathbf{g}^T(k) \cdot \alpha + w(k), \tag{5.6}$$

where

$$\theta = [a(n), b(n), \ldots, a(1), b(1)]^T, \tag{5.7}$$

$$\alpha = [p(n), q(n), \ldots, p(1), q(1)]^T. \tag{5.8}$$

Define a data product matrix

$$\mathbf{S}(k) = \sum_{j=1}^{k} \varphi(j) \varphi^T(j). \tag{5.9}$$

Then the augmented information matrix (AIM) is

$$\mathbf{C}(k) = \mathbf{S}^{-1}(k). \tag{5.10}$$

Decomposing the AIM into its $\mathbf{UDU}^T$ form yields

$$\mathbf{C}(k) = \mathbf{S}^{-1}(k) = \mathbf{U}(k)\mathbf{D}(k)\mathbf{U}^T(k), \tag{5.11}$$

where the parameter matrix $\mathbf{U}(k)$ has the form

$$\mathbf{U}(k) = \begin{bmatrix} 1 & \hat{\alpha}_1^{(1)} & \hat{\theta}_1^{(1)} & \hat{\alpha}_1^{(2)} & \hat{\theta}_1^{(2)} & \cdots & \hat{\alpha}_1^{(n)} & \hat{\theta}_1^{(n)} \\ & 1 & \hat{\theta}_2^{(1)} & \hat{\alpha}_2^{(2)} & \hat{\theta}_2^{(2)} & \cdots & \hat{\alpha}_2^{(n)} & \hat{\theta}_2^{(n)} \\ & & 1 & \hat{\alpha}_3^{(2)} & \hat{\theta}_3^{(3)} & \cdots & \hat{\alpha}_3^{(n)} & \hat{\theta}_3^{(n)} \\ & & & 1 & \hat{\theta}_4^{(2)} & \cdots & \hat{\alpha}_4^{(n)} & \hat{\theta}_4^{(n)} \\ & & & & 1 & \cdots & \hat{\alpha}_5^{(n)} & \hat{\theta}_5^{(n)} \\ & & & & & 1 & \vdots & \vdots \\ & & & & & & 1 & \hat{\theta}_{2n}^{(n)} \\ & & & & & & & 1 \end{bmatrix}, \tag{5.12}$$

and the loss function matrix $\mathbf{D}(k)$ has the form

$$\mathbf{D}(k) = \mathrm{diag}\left[J_f^{(0)}(k) \ J_b^{(1)}(k) \ J_f^{(1)}(k) \ J_b^{(2)}(k) \ J_f^{(2)}(k) \ \cdots \ J_b^{(n)}(k) \ J_f^{(n)}(k) \right],$$
(5.13)

where superscripts "(i)" ($i = 0, 1, 2, \cdots, n$) stand for the model orders. For instance, $\hat{\theta}_1^{(n)}$ is the first parameter of the nth order model parameters. The subscripts "f" and "b" represent the forward and backward paths, respectively.

Define

$$\varphi^T(k)\mathbf{U}(k) = \mathbf{e}^T(k),$$
(5.14)

where

$$\mathbf{e}(k) = \left[e_1(k), e_2(k), \ldots, e_{2n}(k), e_{2n+1}(k) \right]^T.$$
(5.15)

Equation (5.14) is equivalent to the following $(2n+1)$ equations

$$
\begin{aligned}
&0 \Rightarrow z(k-n), \\
&z(k-n), u(k-n) \Rightarrow u(k-n+1), \\
&z(k-n), u(k-n), z(k-n+1) \Rightarrow z(k-n+1), \\
&\vdots \\
&z(k-n), u(k-n), z(k-n+1), \ldots, z(k-1) \Rightarrow u(k-1), \\
&z(k-n), u(k-n), z(k-n+1), \ldots, z(k-1), u(k-1) \Rightarrow z(k),
\end{aligned}
$$
(5.16)

where "$\Rightarrow$" denotes the use of linear combination of the left hand side to fit the right-hand side and $e_i(k)$ is the loss function value of the ith equation [25].

Define the forward path order

$$n_f = \max\{n_a, n_b\}$$
(5.17)

and the backward path order

$$n_b = \max\{n_p, n_q\}.$$
(5.18)

Equation (5.14) describes a set of equalities. The forward path (5.5) is the n_fth odd column of (5.14). The order of (5.5) can be determined by the forward path loss functions $(J_f(k))$, and its parameters, e.g., (5.7) is the corresponding odd column of $U(k)$. Meanwhile, the backward path (5.6) is the n_bth even column of (5.14). The order of (5.6) can be determined by the backward path loss functions $(J_b(k))$, and its parameters, e.g., (5.8) is also the corresponding even column of $U(k)$. For the bivariate process in (5.1), if $Q(z^{-1}) = 0$, and $\mathbf{R}$ is block diagonal, it is causality-free from z to u; if $B(z^{-1}) = 0$ and $\mathbf{R}$ is block diagonal, it is causality-free from u to z.

Therefore the causality relationship between u and z can be anyone of the following four types:

(i) If $B\left(z^{-1}\right) = 0$, $Q\left(z^{-1}\right) = 0$ and $\mathbf{R}$ is block diagonal, then there is no causality between u and z.

(ii) If $B\left(z^{-1}\right) \neq 0$, $Q\left(z^{-1}\right) = 0$ and $\mathbf{R}$ is block diagonal, then u is the cause and z is the effect, but z cannot affect u.

(iii) If $B\left(z^{-1}\right) = 0$, $Q\left(z^{-1}\right) \neq 0$ and $\mathbf{R}$ is block diagonal, then z is the cause and u is the effect, but u cannot affect z.

(iv) Otherwise, u can affect z, and z can also affect u; this is also known as bi-directional causality.

Due to estimation errors, a hypothesis testing should be used to check if the cross-regressive coefficients are close to zero to judge the causality [16].

The above ARMAX process with white noise can be extended to the case with colored noise [18].

This method can be used for identification of the path models between any two variables; in multivariate systems, however, the matrix decomposition is complex because vector autoregressive (VAR) models should be used instead of the above ARMAX model structure.

5.2 Cross-Correlation Analysis

For a multivariate system, it is easy to compute the correlation coefficient between every two variables. This is practical for a preliminary study. However, as mentioned in Chap. 1 correlation does not imply causality; a major difference is that correlation does not show directionality. In this section, a lag-adjusted cross-correlation analysis is introduced to give a similar sense of causality because the concept of temporal direction or lags between the cause and effect is an important aspect of causality.

Assume that x and y are time series of n observations with means μ_x, μ_y and variances σ_x, σ_y, respectively, then the cross-correlation function (CCF) with an assumed lag k is:

$$\phi_{xy}\left(k\right) = \frac{E[(x_i - \mu_x)(y_{i+k} - \mu_y)]}{\sigma_x \sigma_y}, \quad k = -n + 1, \ldots, n - 1. \qquad (5.19)$$

The expectation can be estimated by the sample CCF as:

$$\hat{\phi}_{xy}(k) = \begin{cases} \frac{1}{n-k} \sum_{i=1}^{n-k} (x_i - \mu_x)(y_{i+k} - \mu_y)/s_x s_y, & k \geq 0, \\ \frac{1}{n+k} \sum_{i=1-k}^{n} (x_i - \mu_x)(y_{i+k} - \mu_y)/s_x s_y, & k < 0. \end{cases} \qquad (5.20)$$

where s_x and s_y are sample standard deviations of x and y, respectively.

A value of the CCF is obtained by assuming a certain time delay for one of the time series. Thus the maximum absolute value can be regarded as the real cross-correlation and the corresponding lag as the estimated time delay between these two variables. For mathematical description, one can compute the maximum and minimum values $\phi^{\max} = \max_k \{\phi_{xy}(k), 0\} \geq 0$ and $\phi^{\min} = \min_k \{\phi_{xy}(k), 0\} \leq 0$, and the corresponding arguments $k^{\max}$ and $k^{\min}$. Then the time delay from x to y is:

$$\lambda = \begin{cases} k^{\max}, & \phi^{\max} \geq -\phi^{\min}, \\ k^{\min}, & \phi^{\max} < -\phi^{\min}, \end{cases} \tag{5.21}$$

(corresponding to the maximum absolute value) and the actual time delayed cross-correlation is $\rho = \phi_{xy}(\lambda)$ (between -1 and 1). If λ is less than zero, then it means that the actual delay is from y to x. Thus the sign of λ provides the directionality information between x and y. The sign of ρ corresponds to the sign of the arc in the SDG meaning whether the correlation is positive or negative.

By this definition, ρ is a statistical estimate and is inevitably prone to some uncertainty due to disturbances, noise and the size of data windows. Therefore its value should be judged with care. Even if the two time series are uncorrelated random noise sequences, ρ may still likely be different from zero. Therefore the value of the CCF between two variables should be checked against a threshold. Thus, if the correlation between the two series is very weak, then the effect of the noise will dominate the results. Therefore in correlation analysis, only those values which are significantly larger than a user-defined threshold (e.g., ± 0.2) are considered to be evidence of correlations.

To sum up, based on estimation, the maximum CCF is defined as the time delayed correlation coefficient (or correlation in brief), and the corresponding argument k is defined as the estimate of time delay from x to y, from which we can find the cause and the effect [4]. Of course, correlations are based on statistics under the assumption of linearity, thus they need hypothesis tests to obtain the level of significance. Although the estimates of correlation coefficients are not accurate, the directions are believable for most cases. Although this method is practical and easy to compute, it has many shortcomings, some of which are explained below:

- Nonlinear causal relationship does not necessarily show up in correlation analysis. For example, if y equals the square of x with the time delay of one sample time, where x is a superposition of a sine signal and a white noise, then, based on the time-delayed cross-correlation, this obvious causality cannot be found because all the values are small relative to a threshold, as shown in Fig. 5.1. This can be explained because the true correlation should be zero.
- Correlation simply gives us an estimate of the time delay. The sign of the delay is an estimate of the directionality of the signal flow path. The time delay, however, is only an estimate. In addition, the trend in a time series is ignored, and values at different time instants are regarded as samples of the same random event. Thus the causality obtained by this measure is purely the time delay based on the estimate of the covariance.

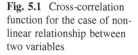

Fig. 5.1 Cross-correlation function for the case of non-linear relationship between two variables

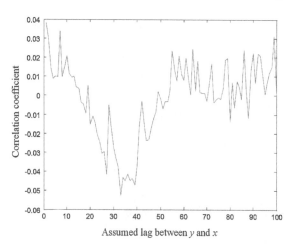

Given all the lag-adjusted cross-correlations between any two variables, a correlation color map (similar to the one proposed in [35]) can be constructed, whose horizontal and vertical coordinates are both variables in the same order and the color of each pixel shows the correlation between the two corresponding variables according to a scaled color bar. This provides an intuitive way to observe correlation, especially for identification of similar groups of variables.

In a system with many variables, pairwise analysis is not enough. Apart from the pair of variables, other related variables can affect the correlation between the two variables. Thus process knowledge should be taken into account to identify the topology in the group of correlated variables. In any case, an apparent correlation should be analyzed no matter whether the causality is direct or indirect.

5.3 Granger Causality Analysis

Regression is a natural way to test the relationship between variables. By taking dynamics into account, the lags in the models reflect causality. A regression of a variable on lagged values of itself is compared with the regression augmented with lagged values of the other variable. If the augmentation is helpful for better regression, then one can conclude that this variable is Granger-caused by the other variable.

This idea comes from [36] who proposed a notion of causality, "*X could be termed as to 'cause' Y if the predictability of Y is improved by incorporating information about X*". However, Wiener's idea lacked the machinery for practical implementation. Later, [14] adopted and formalized this idea in the context of linear regression models. He pointed out that if the incorporation of the past measurements from one time series can reduce the variance of the autoregressive prediction error of a second time series at the present time, then it is said that the first time series has a causal influence on the second one.

Consider two stochastic processes X_t and Y_t. Both of them can be modeled as a vector autoregressive (VAR) process:

$$X_t = \sum_{j=1}^{\infty} a_{1j} X_{t-j} + \varepsilon_{1t}, \quad \text{var}(\varepsilon_{1t}) = \Sigma_1, \tag{5.22}$$

$$Y_t = \sum_{j=1}^{\infty} d_{1j} Y_{t-j} + \eta_{1t}, \quad \text{var}(\eta_{1t}) = \Gamma_1. \tag{5.23}$$

Jointly, they can be represented as

$$X_t = \sum_{j=1}^{\infty} a_{2j} X_{t-j} + \sum_{j=1}^{\infty} b_{2j} Y_{t-j} + \varepsilon_{2t}, \tag{5.24}$$

$$Y_t = \sum_{j=1}^{\infty} c_{2j} X_{t-j} + \sum_{j=1}^{\infty} d_{2j} Y_{t-j} + \eta_{2t}, \tag{5.25}$$

where the noise terms are uncorrelated over time and their contemporaneous covariance matrix is

$$\Sigma = \begin{pmatrix} \Sigma_2 & \Upsilon_2 \\ \Upsilon_2 & \Gamma_2 \end{pmatrix}, \tag{5.26}$$

where $\Sigma_2 = \text{var}(\varepsilon_{2t})$, $\Gamma_2 = \text{var}(\eta_{2t})$, and $\Upsilon_2 = \text{cov}(\varepsilon_{2t}, \eta_{2t})$. If X_t and Y_t are independent, then $\{b_{2j}\}$ and $\{c_{2j}\}$ are zero; thus $\Upsilon_2 = 0$, $\Sigma_1 = \Sigma_2$, and $\Gamma_1 = \Gamma_2$. The total interdependence between X_t and Y_t is defined as

$$F_{X,Y} = \ln \frac{\Sigma_1 \Gamma_1}{|\Sigma|}, \tag{5.27}$$

where $|\ \ |$ represents the determinant of the enclosed matrix. According to this definition, $F_{X,Y} = 0$ when the two time series are independent, and $F_{X,Y} > 0$ when they are not.

Consider (5.22) and (5.24). The value of Σ_1 measures the accuracy of the autoregressive prediction of X_t based on its previous values. The value of Σ_2 represents the accuracy of predicting the present value of X_t based on the previous values of both X_t and Y_t. According to the idea of Granger, if Σ_2 is less than Σ_1, then it is said that Y_t has a causal influence on X_t. This causal influence is defined as follows:

$$F_{Y \to X} = \ln \frac{\Sigma_1}{\Sigma_2}. \tag{5.28}$$

It is noted that, if $F_{Y \to X} = 0$, there is no causal influence from Y_t to X_t, and if $F_{Y \to X} > 0$, there is. Similarly, one can define causal influence from X_t to Y_t as:

$$F_{X \to Y} = \ln \frac{\Gamma_1}{\Gamma_2}. \tag{5.29}$$

It is possible that the interdependence between X_t and Y_t cannot be fully explained by their interactions. The remaining interdependence is captured by Υ_2, the covariance between ε_{2t} and η_{2t}. This interdependence is referred to as an instantaneous causality and is characterized by

$$F_{X \cdot Y} = \ln \frac{\Sigma_2 \Gamma_2}{|\Sigma|}. \tag{5.30}$$

When Υ_2 is zero, $F_{X \cdot Y} = 0$; when Υ_2 is not zero, $F_{X \cdot Y} > 0$.

The above definitions imply that

$$F_{X,Y} = F_{X \to Y} + F_{Y \to X} + F_{X \cdot Y}. \tag{5.31}$$

It is noted that the total interdependence between two time series X_t and Y_t is decomposed into three components: two directional causal influences due to their interaction patterns, and the instantaneous causality due to factors possibly exogenous to the (X, Y) system corresponding to the last term in (5.31).

Granger's concept of causality has received a lot of attention. Indeed, natural time series can also be interpreted in the frequency domain and so it is important to have a spectral representation of causal influence. The major progress in this area was made by Geweke, whose novel decomposition of the multivariate autoregressive process leads to a set of causality measures which have a spectral representation and make the interpretation more straightforward [11, 12].

Consider three processes X_t, Y_t, and Z_t. Suppose that a pairwise analysis reveals a causal influence from Y_t to X_t, in this case one can carry out the following steps to discriminate whether this influence is direct (Fig. 5.2b), or is mediated by Z_t (Fig. 5.2a). Firstly, let the joint autoregressive representations of X_t and Z_t be

$$X_t = \sum_{j=1}^{\infty} a_{3j} X_{t-j} + \sum_{j=1}^{\infty} b_{3j} Z_{t-j} + \varepsilon_{3t}, \tag{5.32}$$

$$Z_t = \sum_{j=1}^{\infty} c_{3j} X_{t-j} + \sum_{j=1}^{\infty} d_{3j} Z_{t-j} + \gamma_{3t}, \tag{5.33}$$

where the covariance matrix of the noise terms is

$$\Sigma_3 = \begin{pmatrix} \Sigma_3 & \Upsilon_3 \\ \Upsilon_3 & \Gamma_3 \end{pmatrix}. \tag{5.34}$$

Fig. 5.2 Two different
patterns of causality

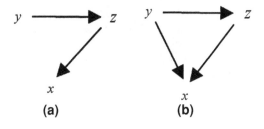

Next consider the joint autoregressive representations of all three processes X_t, Y_t, and Z_t:

$$X_t = \sum_{j=1}^{\infty} a_{4j} X_{t-j} + \sum_{j=1}^{\infty} b_{4j} Y_{t-j} + \sum_{j=1}^{\infty} c_{4j} Z_{t-j} + \varepsilon_{4t}, \tag{5.35}$$

$$Y_t = \sum_{j=1}^{\infty} d_{4j} X_{t-j} + \sum_{j=1}^{\infty} e_{4j} Y_{t-j} + \sum_{j=1}^{\infty} g_{4j} Z_{t-j} + \eta_{4t}, \tag{5.36}$$

$$Z_t = \sum_{j=1}^{\infty} u_{4j} X_{t-j} + \sum_{j=1}^{\infty} v_{4j} Y_{t-j} + \sum_{j=1}^{\infty} w_{4j} Z_{t-j} + \gamma_{4t}, \tag{5.37}$$

where the covariance matrix of the noise terms is

$$\Sigma_4 = \begin{pmatrix} \Sigma_{xx} & \Sigma_{xy} & \Sigma_{xz} \\ \Sigma_{yx} & \Sigma_{yy} & \Sigma_{yz} \\ \Sigma_{zx} & \Sigma_{zy} & \Sigma_{zz} \end{pmatrix}. \tag{5.38}$$

From these two sets of equations, we define the Granger causality from Y_t to X_t conditional on Z_t to be

$$F_{Y \to X|Z} = \ln \frac{\Sigma_3}{\Sigma_{xx}}. \tag{5.39}$$

The intuitive meaning of the definition is that when the causal influence form Y_t to X_t is entirely mediated by Z_t (Fig. 5.2a), $\{b_{4j}\}$ is uniformly zeros, and $\Sigma_3 = \Sigma_{xx}$. Then we have $F_{Y \to X|Z} = 0$, which means that no further improvement in the prediction of X_t can be expected by including past measurements of Y_t. On the other hand, when there is a direct influence from Y_t to X_t (Fig. 5.2b), the inclusion of past measurements of Y_t in addition to that of X_t and Z_t results in better predictions of X_t, leading to $\Sigma_3 > \Sigma_{xx}$ and $F_{Y \to X|Z} > 0$.

In summary, Granger causality is a practical method with acceptable computational burden and has been used in real applications [40]. A Matlab toolbox titled "Causal Connectivity Analysis" is available [32].

The Granger causality method needs a regression model, leading to the following disadvantages of this approach. First, a linear relation between x and y is assumed, which can be quite restrictive. Second, the model accuracy affects the result, especially the predefined model order. There are some extensions of the basic Granger causality concept, such as variants of the Wiener-Granger causality [6], to provide more general forms.

5.4 Directed Transfer Function / Partial Directed Coherence Analysis

A process can also be described in the frequency domain where the energy transfer at every frequency can be shown. Based on this idea, several methods have been developed, such as the directed transfer function (DTF) [19] and the partial directed coherence (PDC) [1]. These quantities DTF and PDC are normalized measures of the total and direct influence respectively between two variables in a multivariate process. Conditioning is conducted to exclude the influence of the confounding variables [13]; this is very important under the multivariate framework [9].

Assume that there are n jointly stationary time series $x_1(k), x_2(k), \cdots, x_n(k)$. As for partial directed coherence (PDC), a jointly stationary multivariate process can be described by an n-dimensional restraint VAR model as shown below, In this case the model order and coefficients $\hat{a}_{ij}(r)(r = 1, \cdots, p)$ are estimated under a certain criterion, such as least squares, based on these n time series:

$$
\begin{cases}
x_1(k) = \sum_r \hat{a}_{11}(r)x_1(k-r) + \sum_r \hat{a}_{12}(r)x_2(k-r) + \cdots + \\
\qquad \sum_r \hat{a}_{1n}(r)x_n(k-r) + \hat{e}_1(k), \\
x_2(k) = \sum_r \hat{a}_{21}(r)x_1(k-r) + \sum_r \hat{a}_{22}(r)x_2(k-r) + \cdots + \\
\qquad \sum_r \hat{a}_{2n}(r)x_n(k-r) + \hat{e}_2(k), \\
\qquad\qquad \cdots\cdots \\
x_n(k) = \sum_r \hat{a}_{n1}(r)x_1(k-r) + \sum_r \hat{a}_{n2}(r)x_2(k-r) + \cdots + \\
\qquad \sum_r \hat{a}_{nn}(r)x_n(k-r) + \hat{e}_n(k).
\end{cases}
\tag{5.40}
$$

Apply Z transform to (5.40), and let $z^{-1} = e^{-j\omega}$, then the frequency response of the process in (5.40) can be written as

$$
\hat{\mathbf{A}}(\omega)\mathbf{X}(\omega) = \mathbf{E}(\omega),
\tag{5.41}
$$

where

$$
\hat{A}_{ij}(\omega) = -\sum_r \hat{a}_{ij}(r)e^{-j\omega r}, \quad \hat{A}_{ii}(\omega) = 1 - \sum_r \hat{a}_{ii}(r)e^{-j\omega r},
\tag{5.42}
$$

$$
\mathbf{X}(\omega) = [x_1(\omega)\ x_2(\omega)\ \dots\ x_n(\omega)]^T,
\tag{5.43}
$$

$$
\mathbf{E}(\omega) = [\hat{e}_1(\omega)\ \hat{e}_2(\omega)\ \dots\ \hat{e}_n(\omega)]^T.
\tag{5.44}
$$

The estimated PDC $\left|\hat{\pi}_{ij}(\omega)\right|$ is defined to reflect the causality from x_j (source node) to x_i (sink node) as [1]:

$$\left|\hat{\pi}_{ij}(\omega)\right| \triangleq \frac{\left|\hat{A}_{ij}(\omega)\right|}{\sqrt{\sum_{i=1}^{n}\left|\hat{A}_{ij}(\omega)\right|^2}}, \tag{5.45}$$

in which the following normalization property holds:

$$0 \leq \left|\hat{\pi}_{ij}(\omega)\right|^2 \leq 1 \text{ and } \sum_{i=1}^{n}\left|\hat{\pi}_{ij}(\omega)\right|^2 = 1, \tag{5.46}$$

and $\left|\hat{\pi}_{ij}(\omega)\right|$ is not zero when x_i is influenced by x_j directly.

In addition, we have

$$\mathbf{X}(z) = \mathbf{A}(z)^{-1}\mathbf{E}(z). \tag{5.47}$$

Letting $\mathbf{H}(z) = \mathbf{A}(z)^{-1}$, we get

$$\mathbf{H}(\omega) = \mathbf{A}(\omega)^{-1} = \begin{bmatrix} h_{11}(\omega) & h_{12}(\omega) & \cdots & h_{1n}(\omega) \\ h_{21}(\omega) & \ddots & \ddots & h_{2n}(\omega) \\ \vdots & \ddots & \ddots & \vdots \\ h_{n1}(\omega) & \cdots & \cdots & h_{nn}(\omega) \end{bmatrix}. \tag{5.48}$$

Assume the covariance matrix of the innovations to be $\mathbf{\Sigma}_e = \mathbf{I}_N$. Then we can define the DTF as:

$$\gamma_{ij}(\omega) = \frac{h_{ij}(\omega)}{\sqrt{\sum_{j=1}^{n}\left|h_{ij}(\omega)\right|^2}}. \tag{5.49}$$

The following normalization property holds:

$$0 \leq \left|\gamma_{ij}\right|^2 \leq 1 \text{ and } \sum_{j=1}^{n}\left|\gamma_{ij}(\omega)\right|^2 = 1. \tag{5.50}$$

Both DTF and PDC can describe the directionality of the effect. DTF measures the total effect of one series on another, which is useful for analyzing fault propagation; but it cannot give further information about whether there exists direct causal influence. Nevertheless, PDC can provide this information, so we could use it to reconstruct the process topology. For a clear visualization, a matrix layout plot is used, as shown in Fig. 3.7.

In addition to a visualization, [29] concluded that PDC is a powerful technique to detect causal influences with respect to Granger causality and derived a significance level for the test. However, there are still some unaddressed problems. Gigi and Tangirala [13] did quantitative analysis on the strength of the causal influences and proved that the total effect, in fact, consists of three components, namely, the direct, indirect, and interference terms. The total effect can be quantified by the DTF, whilst the direct effect is hard to quantify. Zhang et al. [41] also pointed out that the distribution of PDC in the frequency domain remains unaddressed and PDC cannot rank the relative interaction strengths. Baccala and Sameshima [1] asserted that PDC reflects the relative rather than the absolute strength of influence because of the normalization, making the causality given by PDC vulnerable to the number of other signals that are influenced by the same source signals, that is, the causality $x_j \rightarrow x_i$ may change if more (or less) signals are influenced by x_j. To tackle this disadvantage, a renormalized PDC (RPDC) was proposed by Schelter et al. [30], which avoids normalization in its definition.

Let $\boldsymbol{\Sigma}$ and $\mathbf{R}$ denote the covariance matrices of the noise $\mathbf{e}(k) = [e_1(k), \cdots, e_n(k)]^T$ and process $\mathbf{x}(k) = [x_1(k), \cdots, x_n(k)]^T$, respectively. Define $\mathbf{H} \triangleq \mathbf{R}^{-1}$, and

$$\hat{\mathbf{X}}_{ij}(\omega) \triangleq \begin{pmatrix} \mathrm{Re}(\hat{\mathbf{A}}_{ij}(\omega)) \\ \mathrm{Im}(\hat{\mathbf{A}}_{ij}(\omega)) \end{pmatrix}, \tag{5.51}$$

$$\mathbf{V}_{ij}(\omega) \triangleq \sum_{t,l=1}^{p} H_{jj}(t,l)\Sigma_{ii} \begin{pmatrix} \cos(t\omega)\cos(l\omega) & \cos(t\omega)\sin(l\omega) \\ \sin(t\omega)\cos(l\omega) & \sin(t\omega)\sin(l\omega) \end{pmatrix}. \tag{5.52}$$

The index RPDC is defined as [30]

$$\hat{\lambda}_{ij}(\omega) \triangleq N\hat{\mathbf{X}}_{ij}(\omega)'\hat{\mathbf{V}}_{ij}(\omega)^{-1}\hat{\mathbf{X}}_{ij}(\omega), \tag{5.53}$$

where N is the number of data points and $\hat{\mathbf{V}}_{ij}(\omega)$ is the estimate of $\mathbf{V}_{ij}(\omega)$ by substituting estimates $\hat{\mathbf{H}}$ and $\hat{\boldsymbol{\Sigma}}$ for $\mathbf{H}$ and $\boldsymbol{\Sigma}$ in (5.52), in which $\boldsymbol{\Sigma}$ and $\mathbf{R}$ need to be estimated based on $\mathbf{e}(k)$ and $\mathbf{x}(k)$.

Schelter et al. [30] provided the following important proposition: Under the null hypothesis of $\left|A_{ij}(\omega)\right|^2 = 0$, for $p \geq 2$ and $\omega \neq 0 \bmod \pi$, the RPDC $\hat{\lambda}_{ij}(\omega)$ follows an approximate χ^2 distribution with two degrees of freedom as N tends to infinity. When $p = 1$ or $\omega = 0 \bmod \pi$, the RPDC $\hat{\lambda}_{ij}(\omega)$ with $\hat{\mathbf{V}}_{ij}(\omega)^{-1}$ being the generalized inverse of $\hat{\mathbf{V}}_{ij}(\omega)$ is an approximate χ^2 distribution with one degree of freedom as N tends to infinity.

The frequency domain methods have similar advantages as the corresponding time domain methods (Granger causality methods). However, they provide a better insight into the energy transfer description at different frequencies.

5.5 Transfer Entropy Analysis

Both Granger causality and PDC are based on linear model structures that should be identified first; this limits the application. In real cases, if the process cannot be approximated by a linear model, then a more general method should be used. Transfer entropy provides an information-theoretic method to test causality by measuring the reduction of uncertainty. According to information theory, the transfer entropy from stationary time series x to y is defined as [31]

$$t(y|x) = \sum_{y_{i+h}, \mathbf{y}_i, \mathbf{x}_j} p(y_{i+h}, \mathbf{y}_i, \mathbf{x}_j) \cdot \log \frac{p(y_{i+h}|\mathbf{y}_i, \mathbf{x}_j)}{p(y_{i+h}|\mathbf{y}_i)}, \tag{5.54}$$

where p means the complete or conditional probability density function (PDF), $\mathbf{x}_j = [x_j, x_{j-\tau}, \cdots, x_{j-(k-1)\tau}]$, $\mathbf{y}_i = [y_i, y_{i-\tau}, \cdots, y_{i-(l-1)\tau}]$, τ is the sampling period, and h is the prediction horizon. The transfer entropy is a measure of information transfer from x to y by measuring the reduction of uncertainty while assuming predictability. It is defined as the difference between the information about a future observation of y obtained from the simultaneous observation of past values of both x and y, and the information about the future of y obtained from the past values of y alone. It gives a good sense of causality information without having to require the delay information. From experience we can take $\tau = h \leq 4, k = 0$, and $l = 1$ for the initial trial, while the usual way is to test and compare several parameters, especially τ and h. If the transfer entropies in two directions are considered, then $t(x \rightarrow y) = t(y|x) - t(x|y)$ is used as a measure to decide the quantity and direction of information transfer, namely, causality [5].

In (5.54), the PDF can be estimated by histogram or kernel methods [33], which are nonparametric methods, to fit any shape of the distributions. Here the Gaussian kernel method is used because it is more robust than the naive histogram-based method. The Gaussian kernel function is defined as:

$$K(v) = \frac{1}{\sqrt{2\pi}} e^{-\frac{1}{2}v^2}. \tag{5.55}$$

Thus a univariate PDF can be estimated by

$$\hat{p}(x) = \frac{1}{Nh} \sum_{i=1}^{N} K\left(\frac{x - x_i}{h}\right), \tag{5.56}$$

where N is the number of samples, and h is the bandwidth chosen to minimize the mean square error of the PDF estimation calculated by $h = c \cdot \sigma \cdot N^{-1/5}$, where $c = (4/3)^{1/5} \approx 1.06$ according to a normal reference rule-of-thumb [20].

For the multivariate case (q-dimensional), the estimation of PDF is

$$\hat{p}(x_1, x_2, \cdots, x_q) = \frac{1}{N h_1 \cdots h_q} \sum_{i=1}^{N} K\left(\frac{x_1 - x_{i1}}{h_1}\right) K\left(\frac{x_q - x_{iq}}{h_q}\right), \quad (5.57)$$

where $h_s = c \cdot \sigma(x_{is})_{i=1}^{N} \cdot N^{-1/(4+q)}$ for $s = 1, \cdots, q$, and other symbols are the same as in the univariate case.

In order to detect direct and indirect pathways of the information transfer, the definition of a direct transfer entropy (DTE) is introduced as follows.

Since the data analyzed here is uniformly sampled data, as obtained from processes that are continuous, we only consider continuous random variables here. Given three continuous random variables x, y, and z, let them be sampled at time instant i as denoted by $x_i \in [x_{\min}, x_{\max}]$, $y_i \in [y_{\min}, y_{\max}]$, and $z_i \in [z_{\min}, z_{\max}]$ with $i = 1, 2, \ldots, N$, where N is the number of samples. The causal relationships between each pair of these variables can be estimated by calculating transfer entropies [31].

Let y_{i+h_1} denote the value of y at time instant $i + h_1$, that is, h_1 steps in the future from i, and h_1 is referred to as the prediction horizon; $\mathbf{y}_i^{(k_1)} = [y_i, y_{i-\tau_1}, \ldots, y_{i-(k_1-1)\tau_1}]$ and $\mathbf{x}_i^{(l_1)} = [x_i, x_{i-\tau_1}, \ldots, x_{i-(l_1-1)\tau_1}]$ denote embedding vectors with elements from the past values of y and x, respectively (k_1 is the embedding dimension of y and l_1 is the embedding dimension of x); τ_1 is the time interval that allows the scaling in time of the embedded vector, which can be set to be $h_1 = \tau_1 \leq 4$ as a rule of thumb [5]; $f(y_{i+h_1}, \mathbf{y}_i^{(k_1)}, \mathbf{x}_i^{(l_1)})$ denotes the joint PDF, and $f(\cdot|\cdot)$ denotes the conditional PDF, and thus $f(y_{i+h_1}|\mathbf{y}_i^{(k_1)}, \mathbf{x}_i^{(l_1)})$ denotes the conditional PDF of y_{i+h_1} given $\mathbf{y}_i^{(k_1)}$, and $\mathbf{x}_i^{(l_1)}$ and $f(y_{i+h_1}|\mathbf{y}_i^{(k_1)})$ denotes the conditional PDF of y_{i+h_1} given $\mathbf{y}_i^{(k_1)}$. The differential transfer entropy (TE_{diff}) from x to y, for continuous variables, is then calculated as follows:

$$T_{x \to y} = \int f(y_{i+h_1}, \mathbf{y}_i^{(k_1)}, \mathbf{x}_i^{(l_1)}) \cdot \log \frac{f(y_{i+h_1}|\mathbf{y}_i^{(k_1)}, \mathbf{x}_i^{(l_1)})}{f(y_{i+h_1}|\mathbf{y}_i^{(k_1)})} dw, \quad (5.58)$$

where the base of the logarithm is 2 and $\mathbf{w}$ denotes the random vector $[y_{i+h_1}, \mathbf{y}_i^{(k_1)}, \mathbf{x}_i^{(l_1)}]$. By assuming that the elements of $\mathbf{w}$ are $w_1, w_2, \ldots, w_s$, the integral $\int (\cdot) d\mathbf{w}$ denotes $\int_{-\infty}^{\infty} \cdots \int_{-\infty}^{\infty} (\cdot) dw_1 \cdots dw_s$ for simplicity, and similarly for the following integrals.

The transfer entropy from x to y can be understood as the improvement when using the past information of both x and y to predict the future of y compared to only using the past information of y. In other words, the transfer entropy represents the information about a future observation of variable y obtained from the simultaneous observations of past values of both x and y, after discarding the information about the future of y obtained from the past values of y alone, as shown below and Fig. 5.3.

$$T_{x \to y} = H(y_{i+h_1}|\mathbf{y}_i^{(k_1)}) - H(y_{i+h_1}|\mathbf{y}_i^{(k_1)}, \mathbf{x}_i^{(l_1)}), \quad (5.59)$$

where $H(\cdot)$ denotes Shannon entropy.

Fig. 5.3 Physical meaning of transfer entropy

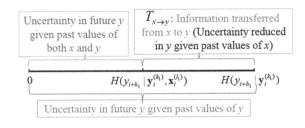

Similarly, the TE$_{\text{diff}}$ from x to z is calculated as follows:

$$T_{x\to z} = \int f(z_{i+h_2}, \mathbf{z}_i^{(m_1)}, \mathbf{x}_i^{(l_2)}) \cdot \log \frac{f(z_{i+h_2}|\mathbf{z}_i^{(m_1)}, \mathbf{x}_i^{(l_2)})}{f(z_{i+h_2}|\mathbf{z}_i^{(m_1)})} d, \qquad (5.60)$$

where h_2 is the prediction horizon, $\mathbf{z}_i^{(m_1)} = [z_i, z_{i-\tau_2}, \ldots, z_{i-(m_1-1)\tau_2}]$ and $\mathbf{x}_i^{(l_2)} = [x_i, x_{i-\tau_2}, \ldots, x_{i-(l_2-1)\tau_2}]$ are embedding vectors with time interval τ_2, and denotes the random vector $[z_{i+h_2}, \mathbf{z}_i^{(m_1)}, \mathbf{x}_i^{(l_2)}]$.

The TE$_{\text{diff}}$ from z to y is calculated as follows:

$$T_{z\to y} = \int f(y_{i+h_3}, \mathbf{y}_i^{(k_2)}, \mathbf{z}_i^{(m_2)}) \cdot \log \frac{f(y_{i+h_3}|\mathbf{y}_i^{(k_2)}, \mathbf{z}_i^{(m_2)})}{f(y_{i+h_3}|\mathbf{y}_i^{(k_2)})} d\mathbf{l}, \qquad (5.61)$$

where h_3 is the prediction horizon, $\mathbf{y}_i^{(k_2)} = [y_i, y_{i-\tau_3}, \ldots, y_{i-(k_2-1)\tau_3}]$ and $\mathbf{z}_i^{(m_2)} = [z_i, z_{i-\tau_3}, \ldots, z_{i-(m_2-1)\tau_3}]$ are embedding vectors with time interval τ_3, and $\mathbf{l}$ denotes the random vector $[y_{i+h_3}, \mathbf{y}_i^{(k_2)}, \mathbf{z}_i^{(m_2)}]$.

If $T_{x\to y}$, $T_{x\to z}$, and $T_{z\to y}$ are all larger than zero, then we conclude that x causes y, x causes z, and z causes y. In this case, we need to distinguish whether the causal influence from x to y is only via the indirect pathway through the intermediate variable z, or in addition to this, there is another direct pathway from x to y. We define a direct causality from x to y as x directly causing y, which means there is a direct information and/or material flow pathway from x to y without any intermediate variables.

In order to detect whether there is a direct causality from x to y, we define a differential direct transfer entropy (DTE$_{\text{diff}}$) from x to y as follows:

$$D_{x\to y} = \int f(y_{i+h}, \mathbf{y}_i^{(k)}, \mathbf{z}_{i+h-h_3}^{(m_2)}, \mathbf{x}_{i+h-h_1}^{(l_1)}) \cdot$$

$$\log \frac{f(y_{i+h}|\mathbf{y}_i^{(k)}, \mathbf{z}_{i+h-h_3}^{(m_2)}, \mathbf{x}_{i+h-h_1}^{(l_1)})}{f(y_{i+h}|\mathbf{y}_i^{(k)}, \mathbf{z}_{i+h-h_3}^{(m_2)})} d\mathbf{v}, \qquad (5.62)$$

where $\mathbf{v}$ denotes the random vector $[y_{i+h}, \mathbf{y}_i^{(k)}, \mathbf{z}_{i+h-h_3}^{(m_2)}, \mathbf{x}_{i+h-h_1}^{(l_1)}]$; the prediction horizon h is set to be $h = \max(h_1, h_3)$; if $h = h_1$, then $\mathbf{y}_i^{(k)} = \mathbf{y}_i^{(k_1)}$, if $h = h_3$, then $\mathbf{y}_i^{(k)} = \mathbf{y}_i^{(k_2)}$; the embedding vector $\mathbf{z}_{i+h-h_3}^{(m_2)} = [z_{i+h-h_3}, z_{i+h-h_3-\tau_3}, \ldots, z_{i+h-h_3-(m_2-1)\tau_3}]$ denotes the past values of z which can provide useful information for predicting the future y at time instant $i + h$, where the embedding dimension m_2 and the time interval τ_3 are determined by (5.61); the embedding vector $\mathbf{x}_{i+h-h_1}^{(l_1)} = [x_{i+h-h_1}, x_{i+h-h_1-\tau_1}, \ldots, x_{i+h-h_1-(l_1-1)\tau_1}]$ denotes the past values of x which can provide useful information to predict the future y at time instant $i + h$, where the embedding dimension l_1 and the time interval τ_1 are determined by (5.58). Note that the parameters in DTE_{diff} are all determined by the calculation of the transfer entropies for consistency.

The DTE_{diff} represents the information about a future observation of y obtained from the simultaneous observation of past values of both x and z, after discarding information about the future y obtained from the past z alone. This can be understood as follows: if the pathway from z to y is cut off, will the history of x still provide some helpful information to predict the future y? Obviously, if this information is non-zero (greater than zero), then there is a direct pathway from x to y. Otherwise there is no direct pathway from x to y, and the causal influence from x to y is all along the indirect pathway via the intermediate variable z.

After the calculation of $D_{x \to y}$, if there is direct causality from x to y, we need to further judge whether the causality from z to y is true or spurious, because it is possible that z is not a cause of y and the spurious causality from z to y is generated by x, i.e., x is the common source of both z and y. As shown in Fig. 5.4, there are still two cases of the information flow pathways between x, y, and z, and the difference is whether there is true and direct causality from z to y.

Thus, DTE_{diff} from z to y needs to be calculated:

$$
D_{z \to y} = \int p(y_{i+h}, \mathbf{y}_i^{(k)}, \mathbf{x}_{i+h-h_1}^{(l_1)}, \mathbf{z}_{i+h-h_3}^{(m_2)}) \cdot
$$
$$
\log \frac{p(y_{i+h} | \mathbf{y}_i^{(k)}, \mathbf{x}_{i+h-h_1}^{(l_1)}, \mathbf{z}_{i+h-h_3}^{(m_2)})}{p(y_{i+h} | \mathbf{y}_i^{(k)}, \mathbf{x}_{i+h-h_1}^{(l_1)})} \, d\mathbf{v}, \tag{5.63}
$$

where the parameters are the same as in (5.62). If $D_{z \to y} > 0$, then there is true and direct causality from z to y, as shown in Fig. 5.4a. Otherwise, the causality from z to y is spurious, which is generated by the common source x, as shown in Fig. 5.4b.

For detailed discussion and calculation methods of this measure, please refer to [8].

Compared to the approach based on cross-correlation, the transfer entropy approach can be applied to more general conditions such as nonlinear relations. In the nonlinear example in Sect. 5.2, causalities cannot be validated based on the cross-correlation term. However, given the lag of 1, the transfer entropies from x to y and vice versa are 0.27 and 0.01 respectively, thus the causality is from x to y, which is consistent with the actual setting.

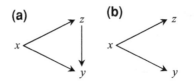

Fig. 5.4 Information flow pathways between x, y, and z with **a** a true and direct causality from z to y and **b** a spurious causality from z to y (meaning that z and y have a common perturbing source, x, and therefore they may appear to be connected or 'correlated' even when they are not connected physically)

Transfer entropy shows the information transfer in each direction. Thus this method provides more insight on causality for complex systems especially for the case with recycles. In Fig. 5.5a, x and y are connected directly via a forward path and a recycle path. In the forward channel from x to y, it is an AR model, i.e., $y(i) + 0.5y(i - 1) = x(i - 5)$; whereas there also exists a feedback channel from y to x, i.e., $x(i) = 1 - 0.5y(i - 1)$. Thus the information transfer lies in both channels. If one is only concerned about the measure $t(x \rightarrow y)$, then the measure $t(x \rightarrow y)$ (in Fig. 5.5b) indicates that the causality is from x to y. However, if the transfer entropies $t(y|x)$ (in Fig. 5.5c) and $t(x|y)$ (in Fig. 5.5d) are studied, then both arcs can be validated.

Barrett and Seth [2] noted that for Gaussian variables, Granger causality and transfer entropy are entirely equivalent. Moreover, [15] extended this equivalence to a weaker condition that is more practical.

To sum up, transfer entropy is a model-free method. There are several parameters to be set by users, which provides some degrees of freedom. However, it has the following main shortcomings. First, it is highly dependent on the estimation of PDFs (although it may take any non-Gaussian forms); thus the computational burden is very high. Second, the time delay cannot be estimated, and the arc signs in SDGs cannot be obtained. Third, the assumption that the time series is stationary does not hold often and thus the noise (may be nonstationary) is often greater than expected These problems affect the computational results.

5.6 Bayesian Network Learning

The above methods are all lag-based. In addition to these, there are some other methods that are based on Bayesian learning. They provide another point of view of causality.

Random phenomena are everywhere in real life, including industrial processes. Due to the existence of random noises, there are stochastic factors that can be studied. The Bayesian network [7] provides a graph with probabilities, where nodes denote fault modes as well as process variables, and arcs denote conditional probabilities.

Fig. 5.5 Transfer entropy
measures. **a** Two variables
and their relations. **b** Transfer entropy from x to y. **c**
Transfer entropy from y to
x. **d** Causality measure based
on transfer entropy which is
the difference between two
transfer entropies

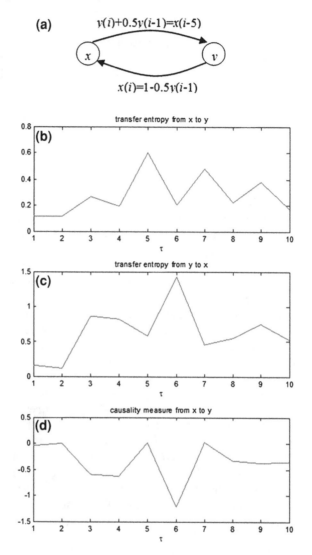

Although the structure remains the same as an ordinary causal graph, both nodes
and arcs mean probabilities. The causality from x to y is described by a conditional
probability $p(y|x)$ [37].

This model is also a general model, although the meaning is different from the
previous ones. It is to be noted that, in industrial processes, dynamics, or time factors, should be included, which is a key feature to capture causality. The traditional
Bayesian network has a fatal limitation that it should be a directed acyclic graph. In
a logical system with no time factors, this assumption makes sense, but in a dynamic

process, cycles are very common. A cyclic causal discovery algorithm has been developed [28] to allow the existence of cycles.

The major limitations of the application of Bayesian networks are: the physical explanation of probabilities is not straightforward, which is sometimes unacceptable by engineers; and the data requirement is hard to meet because one needs sufficient data in all modes, including all fault modes, to build the model.

5.7 Other Methods

In addition to the above methods, there are alternative methods to capture causality between different time series. For example, predictability improvement [3, 10] is another general method with the advantage of requiring a relatively short data set. It computes the reduction of uncertainty of one variable with the help of the other variable. Lungarella et al. and Smith et al. have summarized and compared many methods to capture causality for bivariate series [34] and in a network [21] respectively.

Each of these methods has its own advantages and limitations; they complement each other and no one method is powerful enough to replace the others. Hence we should try different methods to obtain reasonable results. In real applications, one may mainly choose one method but sometimes use other methods to gain additional insights or to validate the results.

Many pairwise data-based methods cannot capture the true causality. If both x and y are driven by a common third variable, sometimes with different lags, one might still find some causality. In fact, there is no causality between these two variables and neither of them can have influence on the other if the third variable does not change. Thus one needs to test all the pairs of variables to obtain their causality measures and then construct the topological structure. The structure should be a mixture of the typical serial structures and parallel structures. Indeed, the topology determination needs additional information beyond pairwise tests.

5.8 Mutual Validation by Process Knowledge and Data

Process knowledge and process data are two means to capture causality information in a process. Neither, however, is sufficient for practical use because of redundancy and errors in the resulting models; we should combine them by mutual validation [38].

5.8.1 Using Process Data to Validate Knowledge Description

Causal networks constructed based on process knowledge are potentially better; ideally, they can cover all possible paths. They also have the ability to avoid indirect

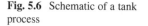

Fig. 5.6 Schematic of a tank process

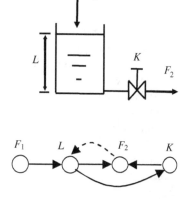

Fig. 5.7 SDG of the tank process

relations by avoiding parallel paths. For example, if A can reach B and B can reach C, then A can reach C. However, if the reachable path from A to C is just realized by the serial connection of the two paths from A to B and from B to C, then the direct arc from A to C can be deleted; otherwise there should exist sound reasons to include different parallel paths to achieve the reachability from A to C, resulting in a necessary arc. The major problem of such causal networks constructed from process knowledge is lack of quantitative information to confirm its reliability. In fact, there are usually redundant and irrelevant arcs that should be deleted.

There are many reasons for this redundancy, the essential one being that the connectivity is merely a necessary condition for causality. While the detailed information can also be obtained to exclude some arcs or choose dominant paths, the effort would be multiplied because quantitative process knowledge is needed in addition to qualitative information. In the modular approach [26], this task is left to the user, although semi-automated; it still relies on experience. Thus we limit the process knowledge of concern to P&IDs, namely, qualitative connectivity information. Under this circumstance, complex structures, such as dividers, headers, or recycle loops, often lead to ambiguous results that are difficult to improve according to qualitative information because the intensity of each arc is unknown. This is an intrinsic problem of the SDG model, although a few researchers have made some improvements [22, 26].

In general, it is difficult to exclude physically broken paths (e.g., valve block), behaviourally uncertain loops (e.g., control loop), or extremely weak (due to attenuation of signals) paths. To validate the causal network, one should resort to process data for quantitative evidence. If there is no data support for reachability, then the causality should be excluded. For the simple tank system shown in Fig. 5.6 (the same as Fig. 1.1), if the data of F_1, F_2, L, and K (take controller output as an alternative) is available and sufficiently excited, then the arcs in Fig. 5.7 can be validated except $F_2 \to L$ because the control determines the value of L.

5.8.2 Using Process Knowledge to Validate Data-Based Relations

Although there are several data-based methods to capture causality, many of them have been developed to find the causality between two variables. Real systems, however, are multivariate, the causality within which is shown as a network with weights based on causality measures between every two variables. Thus a causality matrix is obtained to reflect the magnitude of the causality of each pair of variables, and the direction is determined by the time delay in correlation analysis or the sign of measure in information transfer computation. The topology of the causal network, however, also relies on the propagation relations by screening indirect relations. According to [4], one of the two typical topologies (extreme cases) is generated according to the number of nonzero entries in the first row and above the main diagonal of the causality matrix, while the real topology is the combination of these two topology forms. With the aid of correlation test and directionality test, one can select the evident relations first and construct the network [39]. A method based on correlation is proposed, which introduces the resulting quantitative information obtained by CCF computations:

Step 1: In matrix **P**, select the maximum value in the elements that has not been used and tested.

Step 2: Check the results of correlation test and directionality test. If the correlation value fails to pass the tests, then stop, otherwise go on.

Step 3: Check the result of consistency test for all the variables in the existing arcs. If it fails, then go to Step 5.

Step 4: Add an arc corresponding to this element with an estimated time delay. The sign of the arc is determined by the sign of the element.

Step 5: Go to Step 1.

Nevertheless the resulting network may be incorrect or inconcise without validation by the process knowledge.

One possible way to validate an acceptable causal network is to check the reachability between the two variables for evident causality; this can be realized by searching consistent paths. This treatment, however, cannot exclude indirect relations; the arcs are selected according to their magnitudes of causality. This problem has been discussed in the previous sections. Even if it is possible to identify direct/indirect relations from data analysis, it is not efficient without looking at process knowledge; thus in most cases, one should make good use of process knowledge in the modeling procedure.

For the tank system as shown in Fig. 5.6, based on the data of F_1, F_2, L, and K, one can easily obtain the causal relations from F_1 to F_2, from K to F_2, and from F_1 to K when the level control is in effect. These arcs can be validated by reachability check. If the data set also includes transient response, then other causal relations, such as from F_1 to L, can be detected. The SDG can be constructed accordingly and validated. The structure related to L is slightly different from Fig. 5.7 because of the level control.

5.9 Chapter Summary

We have briefly introduced several data-based causality capturing analysis methods from process data. Pairwise analysis is basic and simple and serves as the foundation for the analysis of multivariate systems. For this purpose, most of the methods, including correlation analysis, Granger causality, and transfer entropy, are suitable and efficient. For multivariate analysis, however, the computational burden is the main bottleneck because the confounding treatment may greatly increase the scale of the problem. Although the multivariate versions of the above methods can deal with it theoretically, its computational burden prevents wide-spread practical application of these methods. Therefore, we should make efforts to reduce the problem scale by the idea of divide and conquer and focus on a smaller set of variables. On the other hand, PDC and DTF can give a good interpretation of the causality among many variables with an acceptable computational burden; thus they do offer a promising alternative.

It is difficult to recommend the most suitable method for a specific problem because each method has its own advantages and disadvantages. Some comparative studies [34] have been conducted and the conclusions therein should be referred to. Some toolboxes that incorporate different methods have been developed, e.g., BioSig (http://biosig.sourceforge.net) and eMVAR (http://www.science.unitn.it/~nollo/research/sigpro/eMVAR.html).

Note that some methods are based on an estimation of model parameters, such as Granger causality and PDC/DTF. Thus the influence of estimation adequacy on the results should be evaluated because it is difficult to identify the model precisely. Due to parsimony, simple model structures are preferred, such as ARMAX models.

Also note that the direct causality here is a relative concept; since the measured process variables are limited, the direct causality analysis is only based on these variables. In other words, even if there are intermediate variables in the connecting pathway between two measured variables, as long as these intermediate variables are unmeasured, we still state that the causality is direct between the pair of measured variables.

The limitation of these methods should also be considered. Some methods require the linearity assumption, which may be fulfilled only approximately in most real cases. In highly nonlinear cases, more general methods should be applied, such as transfer entropy.

References

1. Baccala LA, Sameshima K (2001) Partial directed coherence: a new concept in neural structure determination. Biol Cybern 84(6):463–474
2. Barrett AB, Seth AK (2009) Granger causality and transfer entropy are equivalent for Gaussian variables. Phys Rev Lett 103(238):701

3. Bauer M, Thornhill NF (2005) Measuring cause and effect between process variables. In: Proceedings of the IEEE advanced process control applications for industry workshop, Vancouver, BC, Canada
4. Bauer M, Thornhill NF (2008) A practical method for identifying the propagation path of plant-wide disturbances. J Process Control 18(7–8):707–719
5. Bauer M, Cox JW, Caveness MH, Downs JJ, Thornhill NF (2007) Finding the direction of disturbance propagation in a chemical process using transfer entropy. IEEE Trans Control Syst Technol 15(1):12–21
6. Bressler SL, Seth AK (2011) Wiener-Granger causality: a well established methodology. NeuroImage 58(2):323–329
7. Cowell RG, Dawid AP, Lauritzen SL, Spiegelhalter DJ (1999) Probabilistic networks and expert systems. Springer-Verlag, New York
8. Duan P, Yang F, Chen T, Shah SL (2013) Direct causality detection via the transfer entropy approach. IEEE Trans Control Syst Technol 21(6):2052–2066
9. Faes L, Porta A, Nollo G (2010) Testing frequency-domain causality in multivariate time series. IEEE Trans Biomed Eng 57(8):1897–1906
10. Feldmann U, Bhattacharya J (2004) Predictability improvement as an asymmetrical measure of interdependence in bivariate time series. Int J Bifurcat Chaos 14(2):505–514
11. Geweke J (1982) Measurement of linear dependence and feedback between multiple time series. J Am Stat Assoc 77(378):304–313
12. Geweke J (1984) Measures of conditional linear dependence and feedback between time series. J Am Stat Assoc 79:907–915
13. Gigi S, Tangirala AK (2010) Quantitative analysis of directional strengths in jointly stationary linear multivariate processes. Biol Cybern 103:119–133
14. Granger CWJ (1969) Investigating causal relations by econometric models and cross-spectral methods. Econometrica 37(3):424–438
15. Hlavackova-Schindler K (2011) Equivalence of Granger causality and transfer entropy: a generalization. Appl Math Sci 5(73):3637–3648
16. Jiang B, Yang F, Huang D, Wang W (2012) Extended-AUDI method for simultaneous determination of causality and models from process data. In: Proceedings of American control conference, Washington, DC, pp. 2497–2502
17. Jiang B, Yang F, Jiang Y, Huang D (2012) An extended AUDI algorithm for simultaneous identification of forward and backward paths in closed-loop systems. In: Proceedings of 2012 international symposium on advanced control of chemical processes, Singapore, pp 396–401
18. Jiang B, Yang F, Wang W, Huang D (2014) Simultaneous identification of bi-directional path models based on process data. IEEE Trans Autom Sci Eng. doi:10.1109/TASE.2014.2304536
19. Kaminski M, Blinowska K (1991) A new method of the description of the information flow in the brain structures. Biol Cybern 65(7):203–210
20. Li Q, Racine JS (2007) Nonparametric econometrics: theory and practice. Princeton University Press, Princeton
21. Lungarella M, Ishiguro K, Kuniyoshi Y, Otsu N (2007) Methods for quantifying the causal structure of bivariate time series. Int J Bifurcat Chaos 17(3):903–921
22. Maurya MR, Rengaswamy R, Venkatasubramanian V (2003) A systematic framework for the development and analysis of signed digraphs for chemical processes. 2. control loops and flowsheet analysis. Ind Eng Chem Res 42(20):4811–4827
23. Mayer-Schonberger V, Cukier K (2013) Big data: a revolution that will transform how we live, work, and think. John Murray Publishers, London
24. Niu S, Xiao D, Fisher DG (1990) A recursive algorithm for simultaneous identification of model order and parameters. IEEE Trans Acoust Speech Signal Process 38:884–886
25. Niu S, Fisher DG, Ljung L, Shah SL (1995) A tutorial on multiple model least-squares and augmented ud identification. Technical Report. LiTH-ISY-R-1710, Department of Electrical Engineering, Linkoing University, Linkoping, Sweden
26. Palmer C, Chung PWH (2000) Creating signed directed graph models for process plants. Ind Eng Chem Res 39(20):2548–2558

27. Patel R, Bowman F, Rilling J (2006) A bayesian approach to determining connectivity of the human brain. Hum. Brain Mapp. 27:267–276
28. Richardson T, Spirtes P (2001) Automated discovery of linear feedback models. Computation, Causation, and Causality. MIT Press, New York, pp 1–52
29. Schelter B, Witerhalder M, Michael E, Martin P, Bernhard H, Brigitte G (2005) Testing for directed influences among neural signals using partial directed coherence. J Neurosci Methods 152:210–219
30. Schelter B, Timmer J, Eichler M (2009) Assessing the strength of directed influences among neural signals using renormalized partial directed coherence. J Neurosci Methods 179:121–130
31. Schreiber T (2000) Measuring information transfer. Phys Rev Lett 85(2):461–464
32. Seth A (2010) A MATLAB toolbox for Granger causal connectivity analysis. J Neurosci Methods 186(2):262–273
33. Silverman BW (1986) Density estimation for statistics and data analysis. Chapman and Hall, London
34. Smith SM, Miller KL, Salimi-Khorshidi G, Webster M, Beckmann CF, Nichols TE, Ramsey JD, Woolrich MW (2011) Network modelling methods for FMRI. NeuroImage 54:875–891
35. Tangirala AK, Shah SL, Thornhill NF (2005) PSCMAP: a new tool for plant-wide oscillation detection. J Process Control 15(8):931–941
36. Wiener N (1956) The Theory of Prediction. In: Beckenbach E (ed) Modern mathematics for engineers. McGraw-Hill, New York, pp 165–190
37. Yang F, Xiao D (2006) Approach to fault diagnosis using SDG based on fault revealing time. Proceedings of 6th world congress on intelligent control and automation. Dalian, China, pp 5744–5747
38. Yang F, Xiao D (2012) Progress in root cause and fault propagation analysis of large-scale industrial processes. J Control Sci Eng (Article ID 478373) 2012:1–10. doi:10.1155/2012/478373
39. Yang F, Shah SL, Xiao D (2010) SDG (signed directed graph) based process description and fault propagation analysis for a tailings pumping process. In: Proceedings of 13th IFAC symposium on automation in mining, mineral and metal processing, Cape Town, South Africa
40. Yuan T, Qin SJ (2012) Root cause diagnosis of plant-wide oscillations using Granger causality. In: Proceedings of 8th IFAC international symposium on advanced control of chemical processes, Singapore, pp. 160–165
41. Zhang J, Yang F, Ye H (2013) Quantitative analysis of partial directed coherence in jointly stationary multivariate processes. In: The 23th Chinese process control conference, Guiyang, China

Chapter 6
Case Studies

Abstract Experimental and industrial case studies are provided to show the usefulness of the previously mentioned connectivity and causality analysis techniques for capturing the direction of information flow and diagnosing the likely rootcause(s) of plant-wide oscillations. For an experimental three-tank system, various methods, including adjacency matrix, Granger causality, transfer entropy, and Bayesian network, are applied to capture the connectivity and causality. For the Eastman process with evident oscillation, the above methods are employed to find the faultpropagation pathways and diagnose the root cause of certain disturbance or fault. For a final tailings pump house process, process data and process knowledge are used to build the process topology and to validate each other. Some suggestions for choosing appropriate methods in practice are also given.

Keywords Plant-wide oscillations · Three-tank system · Eastman process · Final tailings pump house · Inference · Causal maps · Data-driven methods · Fault propagation · Validation

In this chapter, experimental and industrial case studies are provided to show the usefulness of the previously mentioned connectivity and causality analysis techniques for capturing the direction of information flow and diagnosing the likely root cause(s) of plant-wide oscillations.

6.1 Three-Tank System

First a 3-tank experiment was conducted, the schematic of which is shown in Fig. 6.1. Water is drawn from a reservoir and pumped to tanks 1 and 2 by a gear pump and a three way valve. The water in tank 2 can flow into tank 3. The water in tanks 1 and

F. Yang et al., *Capturing Connectivity and Causality in Complex Industrial Processes*,
SpringerBriefs in Applied Sciences and Technology,
DOI: 10.1007/978-3-319-05380-6_6, © The Author(s) 2014

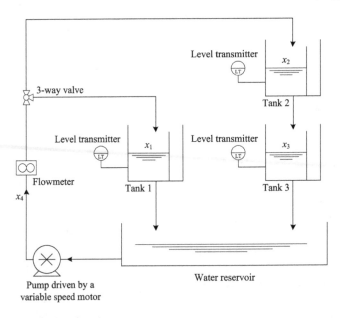

Fig. 6.1 Schematic of the 3-tank system

3 eventually flows into the reservoir. The experiment is conducted under open-loop conditions.

The water levels are measured by level transmitters. We denote the water levels of tanks 1, 2, and 3 by x_1, x_2, and x_3, respectively. The flow rate of the pumped water is measured by a flow meter; we denote this flow rate by x_4. In this experiment, x_4 is set to be a pseudo-random binary sequence (PRBS). The sampled data of 3000 observations is analyzed. Figure 6.2 shows the normalized time trends of the measurements. The sampling time is 1 s.

6.1.1 Adjacency Matrix Method

A directed graph or a digraph represents the structural relationships between discrete objects [4]. The adjacency matrix is a common tool to represent digraphs, which provides an effective way to express process topology. For this 3-tank system, we take each variable x_i as one node i for $i = 1, 2, 3, 4$ and add an edge from x_i (node i) to x_j (node j) if x_i can directly affect x_j without going through any other nodes. Figure 6.3 shows the directed graph of the 3-tank system. For example, the water in tank 2 can flow down into tank 3, if the level of Tank 2, namely, x_2, changes, then the level of tank 3, namely, x_3, will be affected directly. Thus, we add an edge from x_2 to x_3. After a complete analysis of direct interactions between each pair of the nodes, the directed graph of this process is obtained as shown in Fig. 6.3.

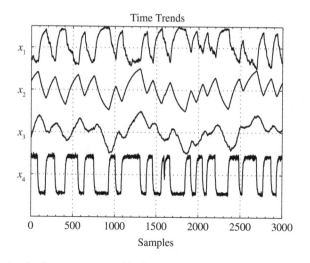

Fig. 6.2 Time trends of measurements of the 3-tank system

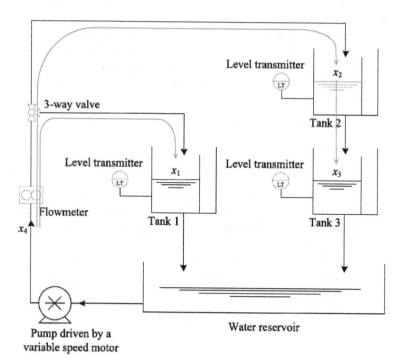

Fig. 6.3 Directed graph of the 3-tank system

Fig. 6.4 Adjacency matrix
and reachability matrix based
on the directed graph for the
3-tank system

(a)

	x_1	x_2	x_3	x_4
x_1	1	0	0	0
x_2	0	1	1	0
x_3	0	0	1	0
x_4	1	1	0	1

Adjacency matrix

(b)

	x_1	x_2	x_3	x_4
x_1	1	0	0	0
x_2	0	1	1	0
x_3	0	0	1	0
x_4	1	1	1	1

Reachability matrix

Based on the directed graph, if there is a directed edge (an arc) from x_i (node i)
to x_j (node j), then the value of (i, j)th entry of the adjacency matrix is set to be
1; otherwise it is 0. We construct the adjacency matrix as shown in Fig. 6.4a. Note
that each node has a direct interaction on itself; thus, all the diagonal elements of
the table are set to be 1. The corresponding reachability matrix is shown in Fig. 6.4b.
From Fig. 6.4b we can see that x_4 can reach all the other nodes while it cannot be
reached by any other nodes.

6.1.2 Granger Causality Method

We first apply time-domain Granger causality to capture the causal relationships
between these four variables. The Akaike information criterion (AIC) is chosen to
determine the model order. For the null hypothesis test, the significance level α is set
to be 0.01, which means that when the p-value is less than 0.01, the null hypothesis
that there is no causality from x_i to x_j is rejected with 99 % confidence level. After
calculation via the Granger causal connectivity analysis (GCCA) toolbox [5], the
causal relationships between the four variables are shown in Fig. 6.5, where the line
with an arrow indicates that there is unidirectional causality from one variable to the
other.

From Fig. 6.5, we can see that x_4 causes x_1 and x_2, and x_2 causes x_3. These causal
relationships are consistent with the information and material flow pathways of the
physical 3-tank system (see Fig. 6.1) since the flow rate of the water out of the pump
decides the water levels of tanks 1 and 2, and the water level of tank 2 affects that of
tank 3. The Granger causality detection method also shows the causal relationship
from x_2 to x_1; the reason for this is that the flow rate of the water out of pump (x_4)

Fig. 6.5 Causal map for the
3-tank system via the Granger
causality method and the
Bayesian network inference
method

Table 6.1 Normalized transfer entropies for the 3-tank system

$NTE^c_{\text{column1}\rightarrow\text{row1}}$	x_1	x_2	x_3	x_4
x_1	NA	0.01	0.19	0.02
x_2	0	NA	0.21	0.02
x_3	0	0	NA	0.02
x_4	0.23	0.21	0	NA

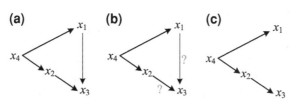

Fig. 6.6 Causal map for the 3-tank system based on **a** and **b** calculation results of TEs which represent the total causality including both direct and indirect/spurious causality; **c** calculation results of DTEs which correctly indicate the direct and true causality

affects water levels in both tanks 1 and 2 (x_1 and x_2), and the spurious causality from x_2 to x_1 is generated by x_4, that is, x_4 is the common source of both x_1 and x_2.

6.1.3 Transfer Entropy Method

The direct transfer entropy (DTE) approach described in [1] is used for causality and direct causality analysis. First the causal relationships between the four variables are detected by calculating the normalized differential transfer entropies (TEs). The calculation results are shown in Table 6.1. Note that the variables listed in column one are the cause variables and the corresponding effect variables appear in the first row.

For the normalized transfer entropies in Table 6.1, we can choose the threshold as 0.05: if the normalized transfer entropy is less than or equal to 0.05, then we conclude that there is almost no causality. We can see that x_1 causes x_3, x_2 causes x_3, and x_4 causes x_1 and x_2. The corresponding causal map is shown in Fig. 6.6a.

Now we need to determine whether the causality between x_1 and x_3 and between x_2 and x_3 is true or spurious, as shown in Fig. 6.6b. To clarify this we first calculate the direct transfer entropy from x_1 to x_3 with intermediate variables x_4 and x_2 and obtain $D^0_{x_1 \to x_3} = 0$, which means that there is no direct information/material flow pathway from x_1 to x_3 and the direct link should be eliminated. Note that we do not need to further detect whether the causality from x_2 to x_3 is true or spurious since there is no intermediate variable or a common cause of both x_2 and x_3 any more after the link from x_1 to x_3 is eliminated. The causality from x_2 to x_3 must be true and direct. The corresponding causal map according to this calculation is shown in Fig. 6.6c, which is consistent with the information and material flow pathways of the physical 3-tank system (see Fig. 6.1).

6.1.4 Bayesian Network Structure Inference Method

The dynamic Bayesian network (BN) inference concept proposed in [10] is applied to capture causality between the four variables. In order to include past information of variables, the lag node order for each variable is chosen to be 3, which means that each variable x_i for $i = 1, 2, 3, 4$ is represented by three lag-compensated nodes: x_i^k, x_i^{k-1}, and x_i^{k-2} representing the current information of x_i at the time instant k and its past information at time instants $k-1, k-2$, respectively. Note that increasing the lag order will increase the computational burden. The larger the order of lags within a certain range, the more accurate the obtained structure is. Here a certain range is similar to the embedding dimension of the embedding vectors with elements from the past values of each variable, which includes all the useful past information of each variable for forecasting other variables. BIC score function and K2 algorithm are chosen to inference the BN structure. Details on the K2 algorithm can be found in [10].

The causal relationships between the four variables via the BN structure inference method are exactly the same as the results via the Granger causality method, as shown in Fig. 6.5. As analyzed above, these causal relationships are consistent with the information and material flow pathways of the physical 3-tank system (see Fig. 6.1).

6.2 Eastman Process

An important application of connectivity and causality analysis is to find the fault propagation pathways and diagnose the root cause of certain disturbance or fault. We use a benchmark industrial data set [2, 7] provided by the Advanced Controls Technology group of Eastman Chemical Company, USA, to illustrate the effectiveness of the commonly used causality detection methods. The Advanced Controls Technology group identified a need to diagnose a common oscillation of about 2 h (about 320 samples/cycle). It is assumed that this common oscillation is probably generated within a certain control loop. The process schematic is shown in Fig. 6.7.

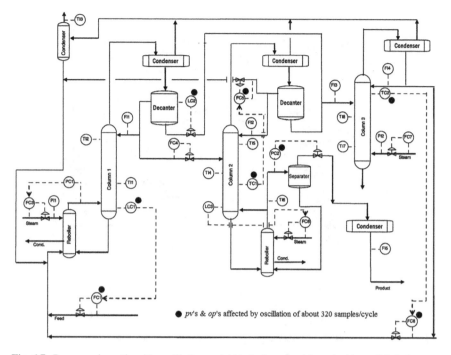

Fig. 6.7 Process schematic with oscillating variables (*pv*'s and *op*'s) marked by *solid circles*

The process contains three distillation columns, two decanters and several recycle streams.

Oscillations are present in the process variables (controlled variables), controller outputs, set points, controller errors (errors between process variables measurements and their set points) or in the measurements from other sensors. The plant-wide oscillation detection and diagnosis methods can be used for any of these time trends [6, 8]. Here we use process variables and controller outputs for root cause analysis; 14 controlled process variables along with 14 PID controller outputs are available. For our study, 5040 sampled observations (from 28 h of data with the sampling interval of 20 s) are analyzed. In this case study, FC, LC, PC and TC represent flow, level, pressure and temperature controllers, respectively. We label the process variable and the controller output by *pv* and *op*, respectively.

Figure 6.8 shows the normalized time trends and power spectra of the 14 process variables (*pv*'s) and Fig. 6.9 shows the normalized time trends and power spectra of the 14 controller outputs (*op*'s). The power spectra indicate the presence of oscillation at the frequency of about 0.003 cycles/sample, corresponding to an approximate period of 2 h. This oscillation propagates throughout the inter-connected units and affects many variables in the process. Thus, our goal is to detect and diagnose the root cause of this oscillation.

For oscillation detection, the spectral envelope method is applied to determine which variables have oscillation at the frequency of 0.0032 cycles/sample. Details

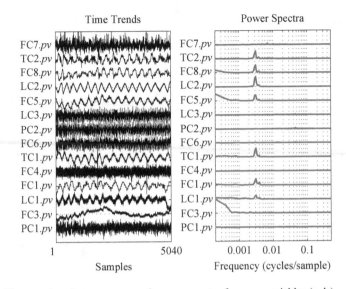

Fig. 6.8 Time trends and power spectra of measurements of process variables (*pv*'s)

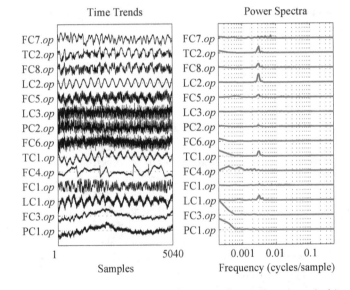

Fig. 6.9 Time trends and power spectra of measurements of controller outputs (*op*'s)

on the spectral envelope method can be found in [2]. Since this book focuses on causality detection and its application to root cause diagnosis, we omit details of oscillation detection and only show the detection result, that is, the *pv*'s and *op*'s of the following eight control loops have common oscillations with 99.9 % confidence level: LC1, FC1, TC1, PC2, FC5, LC2, FC8, and TC2. These control loops with oscillating *pv*'s and *op*'s are marked by solid circles in Fig. 6.7.

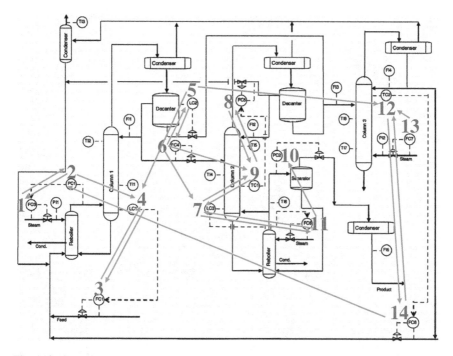

Fig. 6.10 Control loop digraph of the process from Eastman Chemical Company [3]

After detection of the plant-wide oscillation, we show and compare the usefulness of the adjacency matrix method and three process data-based methods including the Granger Causality method, the transfer entropy method, and the Bayesian network inference method for root cause diagnosis of this oscillation.

6.2.1 Adjacency Matrix Method

For the adjacency matrix method, first we need to draw the control loop digraph of the Eastman chemical process as reported in [3]. There are 14 PID controllers, we take each controller as one node and add an edge from node i to node j if $i.op$ can directly affect $j.pv$ without going through controller output of any other node. Figure 6.10 [3] shows the control loop digraph of the Eastman chemical process. For example, node 1 and node 2 are the secondary and the master controllers in a cascade control loop. If the OP of node 1 changes, then the pv of node 2 will be affected directly. Similarly, the op of node 2 has a direct influence on the pv of node 1. Therefore, we say that nodes 1 and 2 have direct interactions between them and we add edges between nodes 1 and 2. Another example is the interaction from node 6 to node 9. If the op of node 6 deviates, the pv of node 9 will be affected directly without going through controller output of any other nodes. Therefore, we add an edge from node

(a)

		1 FC3	2 PC1	3 FC1	4 LC1	5 LC2	6 FC4	7 LC3	8 FC5	9 TC1	10 PC2	11 FC6	12 TC2	13 FC7	14 FC8
1	FC3	1	1	0	0	0	0	0	0	0	0	0	0	0	0
2	PC1	1	1	0	1	0	0	0	0	0	0	0	0	0	0
3	FC1	0	0	1	1	0	0	0	0	0	0	0	0	0	0
4	LC1	0	0	1	1	0	0	0	0	0	0	0	0	0	0
5	LC2	0	0	0	1	1	1	0	0	0	0	0	1	0	0
6	FC4	0	0	0	0	1	1	1	0	1	0	0	0	0	0
7	LC3	0	0	0	0	0	0	1	0	1	0	1	0	0	0
8	FC5	0	0	0	0	0	0	0	1	1	0	0	0	0	0
9	TC1	0	0	0	0	0	0	1	1	1	0	0	0	0	0
10	PC2	0	0	0	0	0	0	0	0	0	1	0	0	0	0
11	FC6	0	0	0	0	0	0	1	0	0	1	1	0	0	0
12	TC2	0	0	0	0	0	0	0	0	0	0	0	1	0	1
13	FC7	0	0	0	0	0	0	0	0	0	0	0	1	1	0
14	FC8	0	1	0	0	0	0	0	0	0	0	0	1	0	1

Adjacency matrix

(b)

		1 FC3	2 PC1	3 FC1	4 LC1	5 LC2	6 FC4	7 LC3	8 FC5	9 TC1	10 PC2	11 FC6	12 TC2	13 FC7	14 FC8
1	FC3	1	1	1	1	0	0	0	0	0	0	0	0	0	0
2	PC1	1	1	1	1	0	0	0	0	0	0	0	0	0	0
3	FC1	0	0	1	1	0	0	0	0	0	0	0	0	0	0
4	LC1	0	0	1	1	0	0	0	0	0	0	0	0	0	0
5	LC2	1	1	1	1	1	1	1	1	1	1	1	1	0	1
6	FC4	1	1	1	1	1	1	1	1	1	1	1	1	0	1
7	LC3	0	0	0	0	0	0	1	1	1	1	1	0	0	0
8	FC5	0	0	0	0	0	0	1	1	1	1	0	0	0	0
9	TC1	0	0	0	0	0	0	1	1	1	1	1	0	0	0
10	PC2	0	0	0	0	0	0	0	0	0	1	0	0	0	0
11	FC6	0	0	0	0	0	0	1	1	1	1	1	0	0	0
12	TC2	1	1	1	1	0	0	0	0	0	0	0	1	0	1
13	FC7	1	1	1	1	0	0	0	0	0	0	0	1	1	1
14	FC8	1	1	1	1	0	0	0	0	0	0	0	1	0	1

Reachability matrix

Fig. 6.11 Adjacency matrix and reachability matrix based on the control loop digraph

6 to node 9. After a complete analysis of direct interactions between each pair of the nodes, the control loop digraph of this process is obtained as shown in Fig. 6.10.

Then, based on the control loop digraph, we construct the adjacency matrix as shown in Fig. 6.11a. If there is an edge from node i to node j, then the (i, j)th entry of the adjacency matrix is assigned a value of 1, otherwise it is assigned a value of 0. Note that the *op* of each node/controller has a direct interaction on the *pv* of itself; thus, all the diagonal elements of the table are set to be 1. The corresponding reachability matrix is shown in Fig. 6.11b. The reachability matrix indicates the influence of a controller on another controller: where '1' denotes a link and '0' indicates no connection. From Fig. 6.11b we can see that nodes 5 and 6 can reach all the other nodes except node 13 and no other nodes can reach them.

As mentioned above, we have already detected the oscillation frequency and isolated eight process variables that have the common oscillation frequency. In Fig. 6.11b, the controllers that have oscillating process variables are highlighted in blue color. Based on the reachability matrix we conclude that: since loops 5 (LC2) and 6 (FC4) can reach all the detected oscillatory loops and they cannot be reached by any other oscillatory loops, the root cause should be either one of these control loops: loop 5 or 6. Based on this conclusion, we can further investigate these two loops and confirm possible root causes including an oscillatory disturbances, or tight tuning of the control loop, or process or valve non-linearity. After further investigation by using the corresponding process data, it has been confirmed that valve stiction in control loop 5 (LC2) was the root cause [2, 7]. We can see that the concepts of adjacency matrix and reachability matrix have successfully suggested potential root causes of plant-wide oscillations.

Note that the adjacency matrix method is a knowledge-based (process connectivity or topology obtained from P&ID) method rather than a data-based method. It can be carried out without using any data. After root cause candidates are obtained via the adjacency matrix method, by using other data-based root cause diagnosis methods, including valve stiction diagnosis methods and nonlinearity test methods, we may confirm the root cause of plant-wide oscillations.

6.2.2 Data Driven Methods

It is assumed that if a variable does not show significant power at the common oscillation frequency, then it does not belong to the group of likely root cause variables [2]. Therefore, we only need to find the information flow pathways among the eight process variables and eight controller outputs that have oscillations at the common frequency. As long as we capture the oscillation propagation pathways between these variables, the possible root causes can be determined. Here the Granger causality method, the transfer entropy method and the Bayesian network inference method are used for causality analysis.

6.2.2.1 Granger Causality Method

The time-domain Granger causality is applied to capture the information flow pathways among the 16 oscillating process variables. The BIC criterion is chosen to determine the model order. For the null hypothesis test, the significance level α is set to be 0.05, which means that when the p-value is less than 0.05, the null hypothesis that there is no causality from x_i to x_j is rejected with 95 % confidence level. After calculation, the causal relationships between the 16 oscillating variables are shown in Fig. 6.12, where a green line with an arrow indicates that there is unidirectional causality from one variable to the other, and a red line connecting two variables with-

Fig. 6.12 Causal map of
16 oscillating variables
via the Granger causality
method. A *green line* with an
arrow indicates unidirectional
causality and a *red line*
connecting two variables
without an *arrow* indicates
bidirectional causality

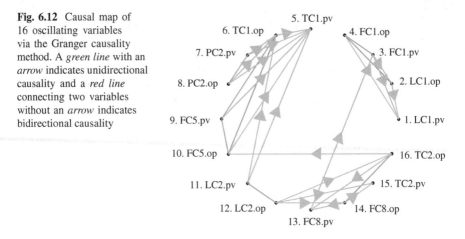

out an arrow indicates there is bidirectional causality (also called causality feedback)
between the two variables.

From the causal map in Fig. 6.12, we can see that there are two control loops, LC2
and PC2, that have causal effects to other loops but do not receive any significant
causal effects from any other loops. Thus, we conclude that control loops LC2 and
PC2 are the likely root cause candidates. Note that the red lines between LC2.*pv*
and LC2.*op*, and between PC2.*pv* and PC2.*op* indicate bidirectional causality which
is generated by the PID feedback control strategy in the control loops. Most of
these causal relationships can be validated by the process schematic. For example,
the bidirectional causality between FC1.op and LC1.op is generated by the cascade
feedback structure between these two loops.

The oscillation propagation pathways obtained from the causal map (see Fig. 6.12)
are shown in Fig. 6.13, where green arrows indicate unidirectional propagation pathways and orange double headed arrows indicate bidirectional propagation pathways.
Note that the bidirectional propagation pathways are generated by the cascade feedback control structure, which are consistent with the physical process. This figure
shows that LC2 can reach all the other loops except PC2, and PC2 can only reach
two loops, i.e., TC1 and FC5. We may conclude that the loop LC2 is the first root
cause candidate and the loop PC2 is the second root cause candidate. Figure 6.13
also shows that the oscillation of loop LC2 propagates to loops TC1, TC2 and FC8
first. By combining with the process schematic shown in Fig. 6.7, this means that the
oscillation of loop LC2 propagates to other loops through material flow pathways
from the left hand side decanter to columns 2 and 3. We can see that these oscillation
propagation pathways in Fig. 6.13 are consistent with the physical process. Since the
root cause of the plant wide oscillation is due to valve stiction in the actuator of the
control loop LC2, the causality analysis via the Granger causality method is effective
in determining the root cause candidate.

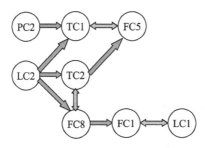

Fig. 6.13 Oscillation propagation pathways obtained via the Granger causality method. *Green arrows* indicate unidirectional propagation pathways and *orange double headed arrows* indicate bidirectional propagation pathways

Fig. 6.14 Causal map of 16 oscillating variables via the transfer entropy method. A *green line* with an *arrow* indicates unidirectional causality and a *red line* connecting two variables without an arrow indicates bidirectional causality

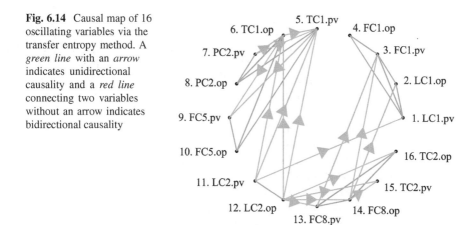

6.2.2.2 Transfer Entropy Method

The normalized differential TE described in [1] is used for causality analysis. We calculated the normalized TE between each pair of the 16 oscillating variables. The threshold for the normalized TE is chosen as 0.05: if the normalized TE is less than or equal to 0.05, then we conclude that there is almost no causality. After calculation, the causal relationships between the 16 oscillating variables are shown in Fig. 6.14.

From the causal map in Fig. 6.14, we can see that control loop LC2 has causal effects on other loops but does not receive any significant causal effects from any other loops. Thus, we conclude that control loop LC2 is likely the root cause candidate. Note that the red line between LC2.*pv* and LC2.*op* indicates bidirectional causality which is generated by the PID feedback controller in the control loop. Most of these causal relationships can be validated by the process schematic.

The oscillation propagation pathways obtained from the causal map (see Fig. 6.14) are shown in Fig. 6.15. This figure shows that the control loop LC2 can reach all the

Fig. 6.15 Oscillation
propagation pathways
obtained via the transfer
entropy method. *Green
arrows* indicate unidirectional
propagation pathways and
*orange double headed
arrows* indicate bidirectional
propagation pathways

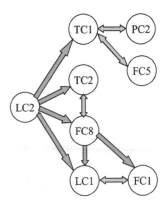

other loops and the oscillation in loop LC2 first propagates to loops TC1, LC1, TC2
and FC8. From Fig. 6.7, we can see that there are direct material flow pathways
from the decanter on the left hand side to columns 1, 2 and 3. Thus, the oscillation
propagation pathways obtained from the transfer entropy method are consistent with
the physical process.

We note that although the conclusion that the control loop LC2 is likely a root
cause candidate is consistent with the Granger causality analysis results, there is an
obvious difference between Fig. 6.12 and Fig. 6.14. Especially, causal relationships
between control loops PC2 and TC1 are found via the transfer entropy method. That
is why PC2 is no longer a root cause candidate according to Fig. 6.14. This conclusion
is consistent with the fact that there was valve stiction in the control loop LC2.

Although there is difference between Figs. 6.12 and 6.14, and between Figs. 6.13
and 6.15, and the conclusions on root cause candidates are not exactly the same, both
the Granger causality method and the transfer entropy method provide effective ways
to capture fault propagation pathways and locate the likely root cause candidates.

6.2.2.3 Bayesian Network Structure Inference Method

The dynamic BN inference concept proposed in [10] is applied to capture causality
between the 16 oscillating variables. In order to include past information of variables,
the lag node order for each variable is chosen to be 3. BIC score function and K2
algorithm [10] are chosen to inference the BN structure. If we include both process
variables and controller outputs, then we need to construct a BN structure with 48
nodes. The computational burden for this BN structure inference is large and probably
we cannot obtain the optimal structure. In order to obtain a more accurate structure
and decrease the computational burden, we only include the eight oscillating process
variables to inference the BN structure.

Figure 6.16 shows the causal relationships between eight oscillating variables.
From the causal map, we can see that there are two variables LC2.*pv* and FC1.*pv* that

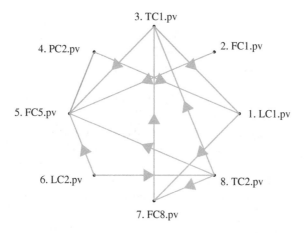

Fig. 6.16 Causal map of the eight oscillating process variables via the BN inference method. A *green line* with an *arrow* indicates unidirectional causality and a *red line* connecting two variables without an *arrow* indicates bidirectional causality

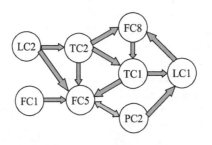

Fig. 6.17 Oscillation propagation pathways obtained via the BN inference method. *Green arrows* indicate unidirectional propagation pathways and *orange double headed arrows* indicate bidirectional propagation pathways

can reach other variables but do not receive causal effects from any other variables. Thus, control loops LC2 and FC1 can be identified as the likely root cause candidates.

The oscillation propagation pathways obtained from the causal map are shown in Fig. 6.17. This figure shows that LC2 can reach all the other loops except FC1, and FC1 can reach all the other loops except LC2 and TC2. Thus, we may conclude that loop LC2 is the first root cause candidate and loop FC1 is the second root cause candidate. Figure 6.17 also shows that the oscillation of loop LC2 propagates to loops TC2 and FC5 first. By combining this information with the process schematic shown in Fig. 6.7, this means that the oscillation of loop LC2 first propagates through material flow pathways from the decanter on left hand side to columns 2 and 3, and then propagates to other loops. We can see that reasonable root cause candidates can also be found based on the analysis results of the BN structure inference method. However, some causal relationships are not captured. For example, the causal relationships between loops LC1 and FC1 are not captured.

In summary, for the three causality analysis methods, although there is difference between the causal maps (see Figs. 6.12, 6.14, and 6.16), and between the oscillation propagation pathways (see Figs. 6.13, 6.15, and 6.17), and the conclusions on root cause candidates are not exactly the same, all three causality detection methods (the transfer entropy method, the Granger causality method, and the Bayesian network inference method) are capable of capturing the fault propagation pathways and locating the likely root cause candidates.

From the above two case studies, we find that causality analysis methods provide an effective way to capture fault propagation pathways. However, the pros and cons of the three data driven methods can be summarized as follows. The major advantages of the Granger causality method are that its theoretical meaning is easy to understand; and its application techniques are well developed. For example, the null hypothesis test of causality is well defined. It is a relatively simple method to implement. A limitation of the application of the Granger causality method is that this method is based on AR models, which is suitable for linear multivariate processes, but the problem of model misspecification may happen and thus the identified AR models may not be convincing. If the model structure is incorrect, then the residuals can hardly give evidence of causality between the signals considered in the models.

For the transfer entropy method, the major advantage is that it can be used for both linear and nonlinear multivariate processes. Its application limitations are: a good parameters determination procedure is needed since the transfer entropy is sensitive to the parameters (e.g., h, k, and l), and the computational burden is large since we need to estimate joint PDFs. Moreover, unlike Granger causality, the distribution of the sample statistic is unknown, rendering significance testing to be difficult without recourse to computationally expensive bootstrap method or the Monte Carlo method by constructing resampling data or surrogate data. Thus, the transfer entropy method is relatively difficult to implement.

For the BN structure inference method, a major advantage is that it can handle the data with a short size, while both the Granger causality method and the transfer entropy method require large data lengths. Disadvantages of the BN structure inference method include the assumption that each observation is independent; this assumption is too strict for industrial process data, and the computational burden is large since we need to estimate the (conditional) PDFs of the data set. Moreover, the results are sensitive to the lags in the nodes and score-based approaches are in general not guaranteed to find the optimal solution. Thus, this method is also relatively difficult to implement.

6.3 Final Tailings Pump House Process

Consider another industrial system, the final tailings pump house (FTPH) process at an oil company in Alberta, Canada, to illustrate the modeling procedure and its validation [9].

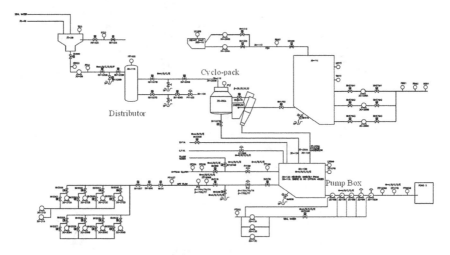

Fig. 6.18 Partial flowsheet of the FTPH process (the texts have been made illegible for confidentiality reasons)

6.3.1 Process Description

The flowsheet for this process is shown in Fig. 6.18, where some texts have been made illegible and the control strategy is omitted for confidentiality reasons. The tailings from upstream are pumped into a distributor and then processed in parallel cyclo-packs and pump boxes, and finally discharged into the ponds. There are five parallel lines from the cyclo-pack downstream, where lines A/B/C/E are structurally identical while line D is distinct. Based on the pressure of the distributor, a prioritization program is implemented on the parallel lines, and Line A is therefore the most important.

The single-loop, cascade, and selective control strategies are applied, including:

- distributor pressure control;
- cyclo-pack pressure and underflow control by adjusting the number of cyclones that are opened;
- gypsum addition flowrate control;
- pump box level control and discharge density control by adjusting cold process water (CPW);
- pump box level control by adjusting pump speed;
- pump box discharge flowrate control by adjusting pump speed, and mature final tailings (MFT) addition flowrate control.

For this process a SDG is constructed from process knowledge and is shown in Fig. 6.19.

For simplicity, six key variables $x_i (i = 1, \cdots, 6)$ in Line A are chosen, which are y10, y16, y18, y21, y28, and y30 respectively, as indicated in Table 6.2. The

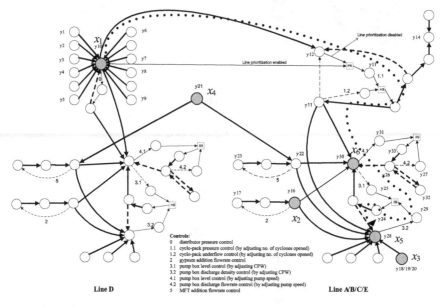

Fig. 6.19 SDG of the FTPH process. Relationships between variables are shown as *thick lines*, and control signal paths are shown as *thin lines*. *Solid* and *dashed lines* show positive (reinforcement) and negative (reduction) causal relations respectively. The *dotted line* is a fault propagation path

Table 6.2 Key process variables in the FTPH process—Line A for causality analysis

Notation	Tag name	Description
x_1	y10	Distributor pressure
x_2	y16	Gypsum addition flow rate
x_3	y18	Gypsum density
x_4	y21	Sludge header pressure
x_5	y28	Sump density
x_6	y30	Sump level

corresponding subsection of the SDG (Fig. 6.20a) is extracted from the SDG in Fig. 6.19; this can be easily done via the reachability check of the SDG model.

6.3.2 Using Process Data to Validate Knowledge Description

Based on the process data of the variables in Line A over a one week long period (with 1 min sampling interval), the correlation color map obtained is displayed in Fig. 6.20c. We use these data and corresponding time delays for SDG validation. For example, the bottom left corner is a cluster of correlated variables. By checking the P&ID, they are found to be associated with the level and density controls in the same pump box.

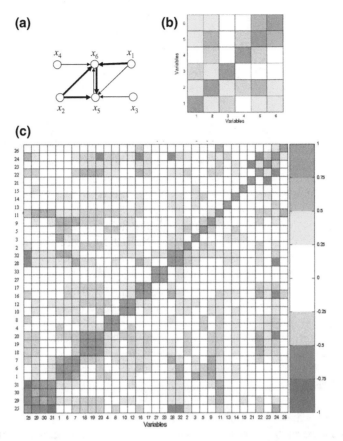

Fig. 6.20 SDG and its validation by cross-correlation analysis of the PVs: **a** subsection of the SDG concerning the six important variables in which the *thick lines* are validated, **b** correlation color map of the six important variables, **c** correlation color map of all the variables in *Line* A

One can focus on the cross-correlations and time delays between the six variables for further analysis. They are shown by the following correlation matrix **P** (comprising of all the correlation coefficients between two variables) and the causality matrix Λ (comprising of all the estimated time delay λ's from one variable to another).

$$
\mathbf{P} = \begin{bmatrix}
1 & 0.28 & -0.28 & -0.18 & 0.39 & 0.41 \\
 & 1 & 0.36 & -0.31 & 0.74 & 0.50 \\
 & & 1 & -0.14 & 0.10 & -0.11 \\
 & & & 1 & -0.25 & -0.24 \\
 & & & & 1 & 0.75 \\
 & & & & & 1
\end{bmatrix}, \tag{6.1}
$$

$$
\Lambda =
\begin{bmatrix}
- & -271 & 220 & 32 & 49 & 5 \\
  & -   & -360 & 2 & 1 & 1 \\
  &   & -   & -357 & 359 & -360 \\
  &   &   & -   & 20 & 60 \\
  &   &   &   & -   & -1 \\
  &   &   &   &   & -
\end{bmatrix}.
\tag{6.2}
$$

For example, the (i, j)th elements of $\mathbf{P}$ and Λ are the correlation and the estimated time delay between variables i and j, respectively. Due to symmetric and anti-symmetric properties of these two matrices, only the elements above the diagonal are computed and shown here. Rather than looking at the correlation matrix in the numerical form, it is better to look at the color coded correlation matrix as shown in Fig. 6.20b which is a portion of Fig. 6.20c.

In (6.1), those values exceeding a pre-set threshold (such as 0.4) can be used to validate some of the arcs in the subsection of the SDG shown in Fig. 6.20a as thick lines. For example, the correlation from x_5 to x_6 is 0.75 and the time delay is -1, so the arc from x_6 to x_5 is validated. Similarly, the arcs from x_2 to x_5, x_2 to x_6, x_1 to x_6, and x_1 to x_5 are also validated. Note that very large time delays do not make sense because they show the ineffectiveness of this measure and the computed correlation is considered invalid.

There are still some arcs that have not been validated. Thus we resort to a more general but more complex measure—transfer entropy. To obtain a rough insight, if we look at the time trends, then we observe that the trends of x_2 and x_4 do not look like the other four variables. This shows that, during this period, the relations associated with x_2 and x_4 are not strong enough as validation. We have to examine more data with sufficient excitement to check if there is a discernible relation between these variables. However, the trends of the other four variables have apparent similarities that should be definitely due to causality. In order to reduce the computational load, we extract from the previous data set only 200 min worth of continuous data shown in Fig. 6.21a.

The transfer entropy measure is used to compute the information transfer between these four variables where τ is assumed to lie between 1 and 10. When τ is 9, the transfer entropies from x_1 to x_6, from x_3 to x_5, and from x_5 to x_6 all reach their individual maximum values: 2.01, 1.48, and 1.41. Thus the bidirectional relationship between x_5 and x_6 is validated, and the reachability from x_3 to x_5 is also validated. They are marked in Fig. 6.21b as thick lines.

By combing the above two methods, we found that most of the arcs have been validated except from x_4 to x_6 and from x_1 to x_5. The former can be explained by the process knowledge because x_6 is the sump level, affected by quite a few variables due to various feeds. The latter is an indirect relation, i.e., x_1 and x_5 are related with the relay of x_6. Since quantitative information is missing in the SDG model, transitive property may be weakened due to the attenuation during the propagation. The time-delayed cross-correlations from x_1 to x_6 and from x_6 to x_5 are 0.41 and 0.75 respectively, but from x_1 to x_5 it is 0.39, less than the above two; moreover, the time delays of the above two are 5 and 1 respectively, while the indirect one is 49,

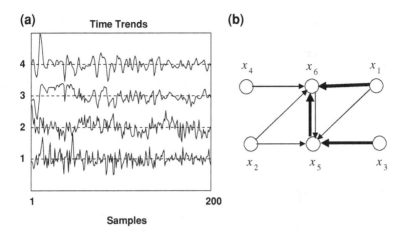

Fig. 6.21 Time trends and validation results by transfer entropy: **a** time trends of four important variables (1: x_1; 2: x_3; 3: x_5; and 4: x_6) in the process, **b** SDG and validated arcs (*thick lines*)

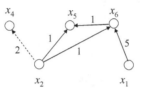

Fig. 6.22 Causal network obtained by correlation analysis. The *solid* and the *dashed lines* are positive and negative correlations respectively. The numbers are the estimated time delays

making the validation result of this arc unacceptable. Therefore, the data analysis is usually used to only validate direct arcs.

6.3.3 Using Process Knowledge to Validate Data-based Relations

Starting from data-based methods, for example correlation analysis, one can obtain the causal network. According to the procedure in Sect. 5.8.2, the maximum 0.75 in (6.1) is selected at the first step, so there is an arc between x_5 and x_6, and the sign is '+'. The corresponding element in Λ is -1 as shown in equation (6.2); thus the direction is from x_6 to x_5. Similarly, the second arc is $x_2 \rightarrow x_5$, and the third one is $x_2 \rightarrow x_6$. For the latter, because the two associated nodes have been used, the consistency test should be undertaken. The time delays associated with the three arcs are 1's that are reliable. Other arcs are built one by one. Note that arc $x_1 \rightarrow x_5$ is ignored because the time delay is 49, which fails to pass the consistency test. The obtained SDG is shown in Fig. 6.22.

However, from the reachability analysis based on the SDG, arc $x_2 \rightarrow x_4$ does not make sense and should be deleted because there is no path connecting them. From process knowledge, they are parameters on different pipes, and in the P&ID of Fig. 6.18, they are independent. One explanation for this link is that there is another cause or upstream unit resulting in the changes to both of them. This incorrect arc may bring out wrong results if the root cause is x_2 because x_4 does not depend on it.

6.3.4 Application of SDGs in Fault Propagation Analysis

Given a validated SDG, the fault propagation can be analyzed qualitatively based on consistent paths. This is important in any conclusive significant HAZOP analysis and also in fault detection and isolation, especially the root cause analysis, where SDGs can help.

In this case, a fault propagation path is shown by the dotted lines in Fig. 6.19 meaning that the pressure change in the distributor (x_1) can affect the parameters in cyclo-packs and pump boxes (x_5 and x_6) in turn. During the HAZOP study, all the consistent paths should be considered and the corresponding consequences should be evaluated. If the domain of influence of one variable is large, the intensity of influence is strong, or the consequence is severe, then some appropriate measures should be taken. In this case study, x_1 is important because it has wide influence on almost all downstream variables; thus the controller on it is well tuned and the line prioritization is implemented to reduce the risk. On the other hand, root cause analysis can be undertaken online; for example, when the variables on this path are showing disturbances, then one can trace immediately the starting point and that may be identified as the root cause of fault propagation. The automation of this procedure will help operators quickly identify the symptom of the abnormal situation.

6.4 Chapter Summary

From the above cases, we have demonstrated the efficacy of description and connectivity and causality capture methods. In real practice, methods should be selected and properly used in terms of the feature and needs of the problem. The methods shown in this chapter are recommended in this sequence, that is, one can first construct the adjacency matrix or SDG based on process knowledge, and then use data-based methods to capture causality; Granger causality can be the first choice due to its simplicity, and transfer entropy can be used for highly nonlinear processes, and Bayesian network and other methods can be used as references or top validate the results of earlier analysis.

References

1. Duan P, Yang F, Chen T, Shah SL (2013) Direct causality detection via the transfer entropy approach. IEEE Trans Control Syst Technol 21(6):2052–2066
2. Jiang H, Choudhury M, Shah SL (2007) Detection and diagnosis of plant-wide oscillations from industrial data using the spectral envelope method. J Process Control 17(2):143–155
3. Jiang H, Patwardhan R, Shah SL (2009) Root cause diagnosis of plant-wide oscillations using the concept of adjacency matrix. J Process Control 19(8):1347–1354
4. Mah RSH (1990) Chemical Process Structures and Information Flows. Butterworth-Heinemann, Oxford
5. Seth A (2010) A MATLAB toolbox for Granger causal connectivity analysis. J Neurosci Methods 186(2):262–273
6. Thornhill NF, Shah SL, Huang B, Vishnubhotla BA (2002) Spectral principal component analysis of dynamic process data. Control Eng Pract 10(8):833–846
7. Thornhill NF, Cox JW, Paulonis MA (2003) Diagnosis of plant-wide oscillation through data-driven analysis and process understanding. Control Eng Pract 11(12):1481–1490
8. Thornhill NF, Huang B, Zhang H (2003) Detection of multiple oscillations in control loops. J Process Control 13(1):91–100
9. Yang F, Shah SL, Xiao D (2012) Signed directed graph based modeling and its validation from process knowledge and process data. Int J Appl Math Comput Sci 22(1):41–53
10. Zou C, Feng J (2009) Granger causality vs. dynamic Bayesian network inference: a comparative study. BMC Bioinformatics 10(122):1–17

Glossary

AE	Algebraic equation
AIC	Akaike information criterion
AIM	Augmented information matrix
AUDI	Augmented upper diagonal identification
CAEX	Computer-aided engineering exchange
CCF	Cross-correlation function
CSTR	Continuous stirred tank reactor
DAE	Differential and algebraic equation
DE	Differential equation
DTE	Direct transfer entropy
DTF	Directed transfer function
FTPH	Final tailings pump house
HAZOP	HAZard and OPerability
IDPUD	Interleaved data pair upper diagonal
MIMO	Multi-input–multi-output
OWL	Web ontology language
P&ID	Piping and instrumentation diagram
PDC	Partial directed coherence
PDF	Probability density function
PFD	Process flow diagram
RDF	Resource description framework
SDG	Signed digraph or signed directed graph
SEM	Structural equation modeling
SISO	Single-input–single-output
XML	eXtensible markup language

F. Yang et al., *Capturing Connectivity and Causality in Complex Industrial Processes*, 91
SpringerBriefs in Applied Sciences and Technology,
DOI: 10.1007/978-3-319-05380-6, © The Author(s) 2014

CPSIA information can be obtained
at www.ICGtesting.com
Printed in the USA
LVHW06s2343250818
588134LV00004B/40/P

9 783319 053790